S0-ADE-392

Nano

Also by Ed Regis

Who Got Einstein's Office?

Great Mambo Chicken and the Transhuman Condition

Nano

the emerging science of nanotechnology: remaking the world—molecule by molecule

Ed Regis

Little, Brown and Company
BOSTON NEW YORK TORONTO LONDON

11/95

First Edition

Library of Congress Cataloging-in-Publication Data

Regis, Edward
Nano : the emerging science of nanotechnology / Ed Regis. — 1st ed.
p. cm.
Includes bibliographical references.
ISBN 0-316-73858-1
1. Nanotechnology. I. Title.
T174.7.R44 1995
620.4 — dc20 94-35378

10 9 8 7 6 5 4 3 2 1

MV-NY

Printed in the United States of America

V

For

Max More
Tom Morrow
Dave Krieger
Romana Machado

> Science advances, funeral by funeral.
>
> — *Anonymous*

Contents

Prologue

Mr. Nano Comes to Washington

On June 26, 1992, at exactly 9:30 in the morning, K. Eric Drexler arrived unaccompanied at room 253 of the Russell Senate Office Building, on Capitol Hill, in Washington. This is a large, white room with a big U-shaped committee table at the front end, a long, green-topped conference table opposite, and, behind it, several rows of seats for the audience. There are about forty well-dressed people sitting there now waiting for events to begin, all of them looking fresh and perky, like a bunch of Catholics, forty bright and shiny faces on this hot June morn.

Drexler's got a briefcase in one arm and a white cardboard box in the other. He's thin, white-faced, and well scrubbed, his dark hair swept straight back. Clad in a dark suit, white shirt, and tie, he looks a bit like Richard Nixon — slightly round-shouldered, an effect completed by his somewhat stiff and formal manner and speech. Not that he's said a word as yet; that will come shortly. In the cardboard box are fifty copies of his prepared statement, a nine-page document headed "Testimony of Dr. K. Eric Drexler on Molecular Nanotechnology before the Senate Committee on Commerce, Science, and Transportation, Subcommittee on Science, Technology, and Space." Drexler has been called to Washington this day from San Francisco, as an expert witness, to tell the country's leaders about his fabulous intellectual creation, his trailblazing new idea, one that, if successfully developed, would stand civilization on its head.

His scheme is to manufacture objects from the molecules up. You'd make things by manipulating individual atoms and molecules, working with them one at a time, positioning them precisely, lining them up one by one, repeatedly, until enough of them accumulated to form a large-scale, usable entity — such as a car or spaceship, for example. All this would be done automatically, effortlessly, without human hands or labor, by a fleet of tiny, invisible robots. These robots, when they were developed, would do all the world's work: people could sit back and enjoy themselves, drinking their mint juleps in peace and quiet, while these wee and unseen machines pieced together their cars and clothes, their food and homes, atom by atom, molecule by molecule.

This was called "nanotechnology." The robots were called "assemblers." Drexler was called "crazy." Or at least that was how some people regarded him the first time they heard about this radical new scheme of his. But here's the subcommittee chairman now entering the room, a bit late.

"There was a vote at nine-thirty this morning, coinciding with our starting time," he explains, apologizing.

This is Al Gore, the man who within the next few days will be announced as Bill Clinton's running mate, and who within five months will be elected vice president. Gore, it turns out, is a big fan of nanotechnology. At any rate he seems au courant with the subject.

SENATOR GORE: When you use the word *nanotechnology*, just so I'm clear in my own mind about this, the first part of that word, *nano-*, is really a measurement word that connotes something that's *real small*, right?

DR. DREXLER: Yes.

SENATOR GORE: All right. Now there seem to me to be three different ways in which the word has been used. *Nanotechnology* has sometimes been used to describe very small etching operations, of the kind you see in the smallest computer chips. Correct?

DR. DREXLER: Yes.

SENATOR GORE: And that's not really what you're talking about. There'd be some overlap at the boundaries, but that's not really what you're talking about. Secondly, there has been an interesting discussion of what might be called micromachines. And sometimes the word *nanotechnology* has been used to describe that whole effort. Correct?

DR. DREXLER: Yes.

SENATOR GORE: And that's not really what you're talking about, either, though again there's some overlap at the boundary. What you're talking about when you use the phrase *molecular nanotechnology* is really a brand-new approach to fabrication, to manufacturing. . . .

The way we make things now, we take some substance in bulk and then whittle down the bulk to the size of the component we need, and then put different components together, and make something. What you're describing with the phrase *molecular nanotechnology* is a completely different approach which rests on the principle that your first building block is the molecule itself. And you're saying that we have all of the basic research breakthroughs that we need to build things one molecule at a time — all we need is the applications of the research necessary to really do it. And you're saying that the advantages of taking a molecular approach are really quite startling.

Really quite startling. That was the truth.

Nanotechnology, Drexler had explained in articles and books, could achieve all manner of wonders. The properties of a given object, after all, were a function of the arrangements of its atoms and molecules. It followed from this that if you could control those arrangements you could control every physical attribute of that object: you'd have "effectively complete control of the structure of matter," as Drexler had often put it.

It was hard to appreciate everything that this meant, so radical a concept was it, but one thing complete control of the structure of matter meant was *complete control of human biology*, and that in turn meant the eradication of disease and aging. Disease, basically, was a molecular phenomenon, a matter of various crucial molecules being out of place. Sickle-cell anemia, for example, was a result of a single specific amino acid being erroneously located in the structure of hemoglobin: where a molecule of glutamic acid should be, a molecule of valine appeared instead. One displaced amino acid and a person could not process oxygen normally! But that could be fixed if you could put the relevant molecules back where they belonged. Aging, likewise, was a case of molecular loss and misplacement, a condition that could be "cured" by putting the right molecules in the right places. With fleets of tiny programmed robots streaming through your body and blood, all kinds of cellular repairs would be possible.

Another thing nanotechnology meant was the elimination of poverty. Drexler's invisible robots would manufacture so many material goods so cheaply that people could have every physical thing they wanted.

A further thing it meant was the abolition of hunger. With nanotechnology you could synthesize food at home, in a box, from the cheapest possible ingredients. You could turn dirt into steak if you wanted to.

That was an idea Drexler came up with in his college days, at MIT in the late 1970s. He thought that once you had the ability to deal with atoms on an individual basis, you could invent this black box — a "meat machine" or "cabinet beast" or something of the sort — that would physically transform common materials into fresh beef. The machine might be about the size and shape of a microwave oven, for example, and it would work the way a microwave oven did, too, more or less. You'd open the door, shovel in a quantity of grass clippings or tree leaves or old bicycle tires or whatever, and then you'd close the door, fiddle with the controls, and sit back to await results. Two hours later, out rolled a wad of fresh beef.

Well, it sounded incredible. But when you thought about it so did the fact that cattle made beef. What materials did *they* have to work with, after all, but grass, air, water, and sunlight? Not one of these things looked remotely like steak. Mix them all together and they looked like mashed grass. Nevertheless, what were cows but walking meat machines? They manufactured beef night and day, on the hoof, automatically, without any human intervention or guidance whatsoever. That they managed to accomplish this miracle while even being picturesque about it, well, how believable was *that* really?

There was a perfectly good explanation as to how it all happened, of course, but it, too, was the story of the serial transformation and rearrangement of molecules. Cattle made beef by breaking the chemical bonds between certain molecules, shuffling those molecules around together with new ones, and establishing other chemical bonds in the patterns characteristic of beef. Beef was a protein, and proteins were combinations of amino acids. Amino acids, in turn, were nothing but molecules of ordinary elements — carbon, hydrogen, oxygen, and nitrogen — bound together in certain stable chemical configurations. Cows made beef by placing the required molecules into the necessary configurations; Drexler's meat machine would do the same thing. The difference was that whereas the cow did its work biologically, by means of enzymes and other organic reactive agents bumping into each other at random in liquids, the meat machine would work *mechanically*, with each separate molecule being pushed into place by a mechanical hand.

The meat machine would be a mechanical cow, a factory at the level of atoms. This was to be understood quite literally: molecules would be stacked on tiny pallets which would move about on tiny tracks. There would be molecular conveyor belts and rollers, vacuum pumps and sorting mills, gears and sprockets and springs and ball bearings. And there would be fleets of molecular manipulator arms — the "assemblers." An assembler would physically grab on to a molecule — taking it from the pallet or conveyor belt or wherever — bring it to the piece of meat under construction, and mechanically force the molecule into position. Billions of such assemblers working in parallel, each of them cycling back and forth millions of times per second, could synthesize chunks of beef that were absolutely indistinguishable from a cow's.

Farfetched, perhaps, but why wouldn't it work? Molecules were molecules: they didn't care how their bonds were established.

But the biggest news of all was that the meat machine was only a special case of a more general nanotechnological building apparatus. The fact was that if Drexler's assemblers could arrange the right molecules the right ways, then they could build you practically anything — they could synthesize just about any product, automatically. Every imaginable miracle device would be within grasp: self-cleaning carpets, for example — "active rugs" whose fibers rippled like cilia, moving dust and dirt off and away; or the gasoline tree, a genetically altered plant that dispensed gasoline instead of tree sap; or the super-duper home shopping network, in which the blueprints for any desired product were sent by optical fiber to your own personal all-purpose household constructor, whose assemblers then manufactured that object at home, for free. Or if not exactly for free, then for a cost that was trifling.

Nanotechnology would be the universal building engine, the molecular cornucopia.

When people were told about nanotechnology and all its magical wonders the first thing they wanted to know was: When will it happen? How many thousands of years will it take? Which, at the Senate hearing, was what Gore asked Drexler.

> Senator Gore: How far off is this stuff, Dr. Drexler? Suppose molecular nanotechnology got the kind of federal and private support that biotechnology got over the last ten years. What kind of advances would you expect to see by 2010, for example?
>
> Dr. Drexler: That kind of question is one of the hardest to answer in this area. I know how to do calculations of the behavior of molecular machinery; I don't know how to do calculations of the rate of progress of a research program where there's a whole series of challenges to be surmounted.

Drexler was by disposition an extremely conservative engineering type who hated to make predictions about human beings and how long it would take them to accomplish a given thing.

Nevertheless, since he was always asked the "When will it happen?" question, he had worked out an answer, and after some hemming and hawing, he gave it.

DR. DREXLER: I commonly answer that fifteen years would not be surprising for major, large-scale applications.

Fifteen years. And if this was to be believed, a rather strange situation was now occurring in the halls of Congress: a scientist brought in as an expert witness was calmly informing the authorities that in the time it took for a newborn babe to reach adolescence, the country would be on the verge of the biggest and most sudden change in its history: physical labor, assembly lines, paychecks would be things of the past; disease and aging would be gone and forgotten; poverty and hunger would be wiped out. And all of it would happen in fifteen years!

But not a word of it ever got out to the press. There were no cover stories in *Time* or *Newsweek*. Although it was open to the public, none of Drexler's testimony was printed in the *New York Times*. He didn't even make it to "All Things Considered." This was puzzling.

Or maybe it wasn't. "We only cover things that actually happen," said a *Time* editor, "not things that are just supposed to happen." In fact, maybe Drexler's whole scheme was nuts after all. Scientists, some of them, had some rather bad things to say about Eric Drexler.

Calvin Quate, professor of electrical engineering at Stanford, said: "I don't think he should be taken seriously. He's too far out."

Phillip Barth, of the Hewlett-Packard Company, said: "The man is a flake."

Nanotechnology itself came off no better.

"This is the kind of thing we see in *Omni* magazine," said Shalom Wind, of the IBM Thomas J. Watson Research Center. "It's more science fiction than it is science."

"It's this basic hand-waving stuff that anyone can do," said Kurt Mislow, a Princeton University chemist. "It's like science fiction, and it turns me off in a major kind of way."

It was science fiction, a common argument went, because atoms couldn't be manipulated as if they were bricks. A whole other set of rules applied to atoms and molecules: the laws of quantum mechanics, which stated that atoms were these indistinct and fuzzy creatures, tiny nebulous balls of energy. You couldn't pin them down or hold them in place, much less maneuver them around like marbles as Mr. Nano wanted to do. Heisenberg's uncertainty principle, the pillar of modern physics, deep-sixed that idea.

Supposing, however, that that could somehow be gotten around — which didn't seem likely — there was still the problem of thermal agitation to contend with. Molecules were always jostling and bouncing and twitching around; they were always in constant motion. How could you build a mechanical device out of parts that never stood still?

And if by some miracle both those difficulties could be avoided or overcome, then you still had to face the fact that radiation or friction or some other atomic complication would attack your little nanomechanism and mangle it beyond belief. So much for Drexler's nano dreams.

The skeptics had a bit of explaining to do, however, when the name of Richard Feynman cropped up, as it invariably did whenever a discussion of nanotechnology lasted for more than five minutes or so. Even Al Gore knew about Feynman.

> SENATOR GORE: The best evidence that the research breakthroughs and the conceptual breakthroughs have long since occurred is that Dr. Richard Feynman made a speech thirty-three years ago in which he essentially outlined the whole field, and even researchers at the cutting edge today were sort of surprised when they went back and read the speech, and found out that the basic concept had been available for a long time.

Drexler never liked to hear this — that Feynman had more or less said it all, way back in the Dark Ages of thirty-three years ago. Drexler always described Feynman's contribution a bit more . . . circumspectly.

Dr. Drexler: Feynman did indeed point in these directions, in a talk in December of 1959, and that has been an inspiration to many people.

The important thing, however, was that Feynman had claimed that working with atoms was entirely feasible. "The principles of physics, as far as I can see," he'd said, "do not speak against the possibility of maneuvering things atom by atom. It is not an attempt to violate any laws; it is something, in principle, that can be done." But if Feynman, the Nobel Prize–winning physicist — the number-two genius, some said, after Einstein — if *Feynman* had said that way back in 1959, then why were the skeptics complaining, years later, that nanotechnology was "science fiction"?

The skeptics had a further bit of explaining to do when in 1989, exactly thirty years after Feynman predicted it, individual atoms were in fact pinned down, moved, and bodily manipulated despite all the obstacles presented by Heisenberg's uncertainty principle, thermal vibration, radiation, and everything else. This feat was performed at the IBM Almaden Research Center, in San Jose,

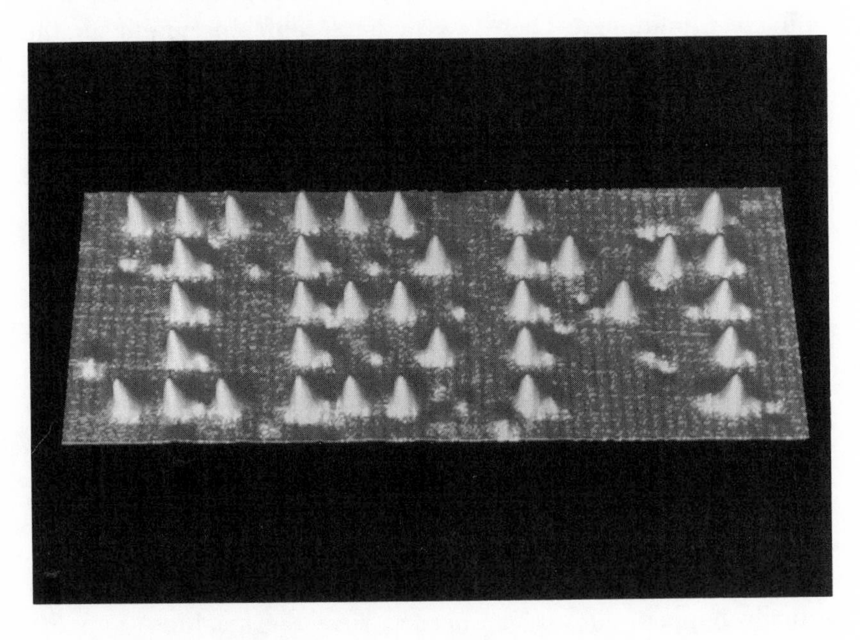

IBM logo spelled out with 35 atoms of xenon. (*IBM Corporation*)

California, when experimenters dragged thirty-five individual atoms of xenon around on a surface until they spelled out the letters IBM.

Suddenly there was a burst of atomic-level creativity in laboratories all over the United States, Germany, and Japan, as hands-on researchers experienced an urgent experimental need to do things like write their names out in atoms, spell the word *Peace* in sulfur molecules, and draw sketches of Albert Einstein in a medium of mixed ions — all of which were accomplished within the next few months.

This was primitive stuff, admittedly, compared to what Eric Drexler was talking about. Still, it was clear that things were beginning to happen down there in the atomic depths. And it was clear that Feynman, at least, had been right all along.

Later that day, after he finished his Senate testimony, Drexler was driven to the Executive Office Building, next door to the White House, where he was to give a briefing to the president's Office of Science and Technology Policy, the OSTP.

This took place in EOB room 476, a small, greenish room overlooking Pennsylvania Avenue, and cooled by a single window-unit air conditioner. There was a long conference table in the middle, at the head of which there was a lectern, an overhead projector, a television set, and a VCR — this for the videotape that Drexler had brought with him.

At about 3:40 P.M. a dozen or so members of the president's science team filed into the room, sat down, and stared in silence as Drexler launched into his hour-long "strategic implications" talk.

Nanotechnology, it turned out, could build more than almost-free cars, houses, sailboats, and rocket ships. It could also build weapons. In fact, it could build some rather creepy weapons — tiny little invisible invaders, ones that would never show up on any radar screen yet invented. They'd be about the size of small bacteria, although slightly more deadly: they could be programmed by some enemy power — or worse, a terrorist group, supposing they managed to get their hands on any — to slip over the border on a gust of wind, enter your body, and turn your bones to slime. This could actually happen. All anyone needed were the machines and the programming.

"It's very hard to picture a future scenario in which we do not have these technologies," Drexler told the president's science crew.

And then he ran his videotape — the one that showed the workings of his molecular planetary gear, in motion and in color.

Planetary gears were these handy little devices for taking the rotation of an inner shaft (the "sun" gear) and converting it to the slower rotations of a bunch of smaller shafts ("planet" gears) that surrounded the inner shaft like planets orbiting the sun. In the Big World such gears were used in automobile drive trains, aircraft engine propeller hubs, and so on. Drexler, together with Ralph Merkle, a full-time nanotechnology researcher at Xerox PARC — the Palo Alto Research Center — had designed a planetary gear at the level of atoms.

There had been a lot of trial and error involved, and some of the earlier designs hadn't worked too well. When they were put through tests on molecular modeling software, things came unglued. Gears slipped out of their housings. Molecular rings exploded like firecrackers. Atomic wreckage flew hither and yon.

But then there was a design that worked perfectly, a gear system that was made out of 3,557 individual atoms — precisely that many, not one more and not one less. Every separate atom was placed just so. All the chemical bonds were correct. And the simulated device worked like a charm. It was up there right now on the TV screen, churning away like an eggbeater.

And there sat the president's science team, eyeballs twitching, as Eric and Ralph's planetary gear vibrated and throbbed and whirled on its bearings.

Is this the wave of the future? Do we really have to worry about this *stuff now?*

Some of those present thought so. "We take it very, very seriously," said Karl Erb, an OSTP member who attended the briefing.

Drexler was back in Washington again a few months later, this time for a private meeting at the Pentagon with Admiral David E. Jeremiah, vice chairman of the Joint Chiefs of Staff. Jeremiah was appalled, given its obvious weapons applications, that so very few military people had ever heard of nanotechnology. Basically, no one had.

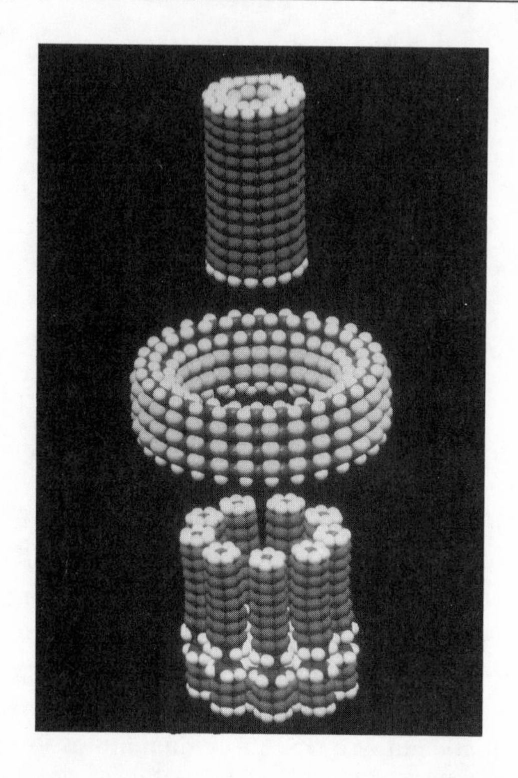

Molecular planetary gear. (*K. Eric Drexler and Ralph C. Merkle*)

"While speaking to a group of senior naval officers last week," he related, "I stressed the need to invest in nanotechnology. After the session, I learned that I had won the prize: not one person there knew what it was. Yet that technology may be the gateway to incredible new systems within a decade."

So Eric came in and met with Admiral Jeremiah — Ralph was there, too, this time — and again he ran through his "strategic implications" talk, the same one he'd given at the White House briefing. Sooner or later somebody's going to develop the technology, he told the admiral. Japan, for one, had already started to do some work; the United States could not afford to fall behind on this sort of thing, et cetera and so forth. Jeremiah, for his part, could only agree: "We want R&D in things like nanotechnology to continue to keep us ahead of potential enemies."

Some people in official Washington, then, were taking Drexler and his creation seriously enough.

And why not? If the pace and direction of scientific research were any indication, nanotechnology truly was the wave of the future. Following their IBM atomic logo, some of the same researchers had gone on to produce an atomic switch, a molecular device in which a single atom was made to toggle back and forth between two positions, on command, just exactly like a Big World electrical switch. "This switch is a prototype of a new class of potentially very small electronic devices," the inventors said.

Chemists had a field day with "buckyballs," tiny soccer-ball-like molecules composed of exactly sixty carbon atoms, and with "nanotubes," elongated cylinders just a few atoms across. These structures, collectively called "fullerenes," were envisioned as building blocks of a whole new race of molecular devices. "All these different fullerenes are like a marvelous construction-toy set," said one researcher. "We can think of a million different ways to put them together to make all kinds of objects that nobody has ever tried before."

Other experimenters were building molecular machines of the most surprising variety. In 1990 J. Fraser Stoddart, a British chemist, put together "a molecular train set," an actual working railroad car made out of no more than a few dozen atoms. Despite its being absolutely invisible to the naked eye, or even in the strongest light microscopes, nuclear magnetic resonance spectroscopy showed that Stoddart's train worked just like its Big World counterpart, traveling along molecular "tracks" and stopping, in succession, at four distinct molecular "stations."

"At room temperature," Stoddart said, "the molecular 'train' is moving three hundred times a second from one 'station' to the next."

He introduced a second train onto the same track — his "circle line," as he called it — and in short order had both of them racing around the circuit in tandem. "The traveling time between 'stations' is hardly altered, with these molecular 'trains' managing to keep a free 'station' between them. Crashes do not happen!"

Well. With these tiny wonders coming up from the labs — built by experimenters for "research" purposes (which was to say, mainly for the fun of it) — and with Eric Drexler's Master Molecular Plan coming down from on high, how could it be anything more than a matter of time before plan and experiment met in the middle, yielding up Drexler's assemblers, his meat machine, his universal molecular constructing engine, and every other imaginable marvel?

As for his timetable, the fifteen-year figure he always mentioned, this, too, was not out of bounds — not if the history of molecular biology was any guide. The structure of DNA had been deciphered in 1953, by Watson and Crick: that was the first time it was known what the DNA molecule actually looked like. But it was only twenty years later, in the early 1970s, that experimenters were able to manipulate those same molecules, cutting them at will, splicing two different DNA strands together, recombining them according to plan, ushering in the much-heralded era of recombinant DNA, genetic engineering, gene therapy, and all the rest.

A few years later and experimenters could actually create, in the lab, their own preprogrammed, predesigned DNA. These were actual DNA molecules — quite genuine in every respect — although never to be found in nature. They'd been made intentionally and in the test tube, instead of naturally by the biological cells.

And a few years after that, the "gene machine" was

invented — a device that made synthetic DNA automatically. Also known as an "automated DNA synthesizer," it fit neatly on an average desktop and could be programmed by anyone who knew how to type. You keyed in the desired base-pair sequences, let the machine run, and in a matter of hours your designer DNA molecules dropped into a bottle, ready for use.

Soon you could look in the classified sections of magazines such as *Science* and *Nature* and read ads that said in their entirety:

QUALITY DNA
Services from $4/ to $9/base
Call Midland
1-800-247-8766

CUSTOM DNA SYNTHESIS
$3.50/base
No set-up fee, shipped in 24 hours.
Keystone Laboratories, Inc.
1-800-788-4 DNA

From discovery of DNA's molecular structure to Federal Express overnight delivery of custom-made, designer DNA molecules — and all in the space of forty years. Why wouldn't the nano era come along at the same speed, in the same headlong rush, if not faster? Technology, after all, had improved a wee bit in the interim. How long could it be before molecular train sets became molecular assemblers? Before gene machines became meat machines?

Not very long, in Drexler's view. It was the ultimate destiny of an intelligent species to accomplish these feats: to move molecules, to make atomically perfect devices, to gain complete control of the structure of matter. It was where science and technology were already headed.

The question was, what would happen, after fifteen years, when the nano revolution finally arrived? Some rather bad things, quite possibly. The thing about nanotechnology was the absolutely incredible power it gave you. Humankind, unfortunately, did not have an especially sterling record when it came to the wise use of power, whether physical or political. The Inquisition, the Third Reich, the atom bomb, and so on — these did not make for a pretty picture.

But nanotechnology dwarfed all those things. It was the Maximum Technology, the final victory over recalcitrant nature. Even if "nano-accidents" could be prevented or contained — molecular machines spinning out of control and eating up Chicago — there was the far worse problem of abuse to worry about, the deliberate misuse of these machines, whether by individuals or governments. The things a backyard bodybuilder could do with some of Drexler's little molecular helpers were horrible to contemplate: steroids were child's play by comparison. You'd have your gigantically overbuilt human bodies, your four-wheel-drive humanoids with their jacked-up muscles, their oversize penises, and God only knew what else.

And as for the possibility of complete control of the structure of matter falling into the wrong hands — such as those that ran some of the more rustic Middle East fatherlands, for example — well, it hardly bore thinking about.

"I expect a big level of freak-out on this," said Stewart Brand, the runic corporate fortune-teller, at a nanotechnology conference. "I expect to see a conference on *stopping* nanotechnology. You could have a nanotechnological Chernobyl. Nanotechnology could become a new abortion issue.

"You can't really get ready for it," he added. "What you're ready for won't happen; what you're not ready for will."

Drexler's greatest worry, indeed, was that nanotechnology would arrive before people were ready for it — before they'd even heard of it. And not very many people had heard of it.

But the prospect was real enough. Nanotechnology "sets us on the threshold of genuinely significant changes," Marvin Minsky, the MIT Renaissance man, had said. "Nanotechnology could have more effect on our material existence than those last two great inventions in that domain — the replacement of sticks and stones by metals and cements, and the harnessing of electricity."

Nanotechnology was already on the horizon, where it was looming forth in all its threat and glory. It might take fifteen years for it to get here, or it might take more. But it was out there and it was approaching — the single biggest technological advance in the history of the species.

And no one knew it.

part i

Limits

1

The *kT* Irony

How could the nano dream *not* seem like science fiction? Atoms were such tiny creatures that you couldn't foresee "working with them" by any known or imaginable means. They seemed to be out of reach, out of touch, off in another realm. Which was rather odd, considering that people were themselves made of atoms, as, of course, was practically everything else in the universe. Nevertheless, the idea of actually reaching out and manipulating these things, well — it was about as believable as holding on to the air. Atoms were so small, so mysterious, so fundamentally alien.

They were far less real than the planet Jupiter. At least you could go out in the backyard and actually see Jupiter: you were convinced that it was really there. You could see pictures of the giant planet, with its patchy and swirling atmosphere, its cloud decks, its Great Red Spot, its various and sundry moons. Jupiter had a top and a bottom; it had shape, form, mass, and definition. It had color. You could imagine going out there on a speeding spaceship and personally inspecting planet Jupiter.

But an atom? That was quite a different story. It wasn't easy to imagine meeting one face-to-face. What would it look like? What color would it be? Would it be rotating or stationary? What was the top and what was the bottom? And how could you tell the difference?

Everyone knew that atoms existed, of course — or at least they thought they knew. But in truth the situation was more like the case of the earth-sun business: everyone "knew" that the earth orbited the sun instead of vice versa, but ask the man on the street what the proof was for this modest proposition, and most probably he wouldn't have the faintest notion. Ask that same person what proof there was that atoms existed, and, . . . well, . . .

As it was, atoms were not definitively proved to exist until the early years of the twentieth century, which was remarkable in view of the fact that atoms had been thought to exist from the time of the ancient Greeks onward. Leucippus and Democritus, the two philosophers who first came up with the idea, advanced the notion of atoms to explain the existence of change, especially the transformation of one substance into another — wood into ashes, for example. Wood and ashes had nothing visible in common, so they must have something invisible in common, and what else could those invisible things be but tiny bits of matter, uncuttable little articles too small to be seen?

That, anyway, was the theory, speculative though it was. The atomic theory began to take on the aspect of a serious empirical hypothesis in 1799, when the French chemist Joseph-Louis Proust noticed some weird ratios cropping up in the makeup of certain chemical compounds he was working with. The constituent elements of those compounds, he noticed, combined only in ratios that were exact multiples of whole numbers. When copper, oxygen, and carbon were mixed to form copper carbonate, for example, they fitted together only in the discrete, whole-number ratios of 5 to 4 to 1, respectively. The proportions were always absolutely uniform and exact, never approximate or fractional — never 5.3 to 4.6 to 1.3, or anything else. Copper carbonate was always just 5 units of copper plus 4 units of oxygen plus 1 unit of carbon.

How could that be true unless, way down at the bottom, the elements existed in the form of tiny units of matter — the "atoms" of yore — which fitted together only as indivisible wholes?

Anyway, that was the theory. So extremely logical was this theory that in the early 1800s, the British chemist John Dalton published a massive three-volume work, *A New System of Chemical*

Philosophy, in which he explained it all in great and arcane detail. He even drew pictures of atoms — tiny circular figures stuck together in various patterns — illustrating how the wee creatures connected up to form the different molecules.

It was hard not to accept the reality of atoms when confronted with Proust's law of definite proportions plus Dalton's entirely convincing molecular portraits. Nevertheless, many scientists, including a number of working chemists, doubted their actual existence, and for the good and sufficient reason that the alleged "atoms" were totally unobservable, far too small to be seen with a magnifying glass, microscope, or by any other means. Unobservable entities were the bane of science, to be dismissed as occult influences.

Thus Jean-Baptiste Dumas, the chemist, said in 1826: "If I were master of the situation I would efface the word atom from Science, persuaded that it goes further than experience, and that, in chemistry, we should never go further than experience."

An atom, said chemist William Prout, was no more than a useful fiction, "a conventional artifice, exceedingly convenient for many purposes but which does not represent nature."

It was a "mystical conception," said Pierre Marcelin Berthelot in 1877. "Who has ever seen a gas molecule or an atom?"

And later still, Ernst Mach, a physicist, asked the same question — "Have you ever seen one?" — whenever his colleagues spoke of those mythological beasts they called "atoms."

"Atoms cannot be perceived by the senses," said Mach. "They can never be seen or touched, and exist only in our imagination. They are things of thought."

Convenient fictions. Mystical conceptions. Things of thought.

That was the situation right down through the opening years of the twentieth century. There were lots of mythical entities banging around in chemistry and physics. There were lots of theories. But even at the time the Wright brothers flew, in 1903, there were still no atoms.

There was, however, Brownian motion.

Robert Brown first observed "Brownian motion" in June 1827,

as he was looking at some pollen grains — tiny particles only about one five-thousandth of an inch across — through a microscope. A rather ample chap with a grim and forbidding expression, he was fifty-four years old at the time, and was generally recognized as one of the world's great botanists. Charles Darwin, who knew him, recalled Brown as "chiefly remarkable for the minuteness of his observations and their perfect accuracy. He never propounded to me any large scientific views in biology. His knowledge was extraordinarily great, and much died with him, owing to his excessive fear of ever making a mistake."

Brown was examining pollen grains of *Clarkia pulchella*, a bright flowering herb native to western North America, when something radically new appeared before him in the microscope.

"While examining the form of these particles immersed in water," he reported, "I observed many of them very evidently in motion. . . . These motions were such as to satisfy me, after frequently repeated observation, that they arose neither from currents in the fluid, nor from its gradual evaporation, but belonged to the particle itself."

This was, as he thought, "peculiar." Still, in the wake of Leeuwenhoek's "animalcules" — protozoa and bacteria swimming about through water — Brown's observations were hardly inexplicable: water, after all, was teeming with living organisms that could be seen only through a microscope. Pollen grains were living things too; why shouldn't they be moving around?

Just to make sure that this was what he was seeing — some microscopic sign of life — Brown decided to kill off the pollen, after which, he was sure, all that herky-jerky motion would stop. So he drenched the grains in alcohol, dried them out, scattered them in water, and put them again under the lens.

Whereupon he saw that the things were still moving. This, he reports, was a "very unexpected fact."

So he decided to see what would happen with previously living organic matter that was now quite certainly dead. He took some fossilized wood, pulverized it, and sprinkled the bits in water — and right away those ancient and inanimate objects somehow managed to move.

He then experimented with "rocks of all ages, including those in which organic remains had never been found." One of those rocks, he reported, was "a fragment of the Sphinx." Nevertheless, the motions of these "active molecules," as he called them, continued.

It now became a personal challenge to find a material that did not harbor these active molecules. Brown experimented with just about every substance he could think of: "travertine, stalactites, lava, obsidian, pumice, volcanic ashes, and meteorites from various localities." But all of them contained the dreaded and puzzling moving objects. "In a word, in every mineral which I could reduce to a powder sufficiently fine to be temporarily suspended in water, I found these molecules more or less copiously."

Brown was by now a bit concerned — as who wouldn't have been? Here were motions that could only be manifestations of primitive forms of life . . . but there could be no life in the rocks, ashes, and other inert stuff he was looking at.

He worked on the problem for three months, and never got anywhere. Finally, in desperation, he began throwing his samples in the fire.

Into the flames with ye molecules and be damned!

But that didn't stop them either. "In all these bodies so heated, quenched in water, and immediately submitted to examination, the molecules were found, and in as evident motion as those obtained from the same substances before burning."

This, indeed, was a mystery. If only he could figure out the source of the motion. But he couldn't. "I shall not at present enter into additional details, nor shall I hazard any conjectures whatever respecting these molecules, which appear to be of such general existence in inorganic as well as in organic bodies."

Here was an observation, but no theory to account for it.

By the turn of the twentieth century, Brownian motion had all but been forgotten by physicists. Some of Brown's contemporaries, indeed, had ridiculed his "observations" at the time.

"I don't believe a word on't," said Oxford professor Henry

Conybeare. "I would only add one supposition more; that these molecules are inhabited."

But in 1905, someone finally took Brownian motion seriously enough to come up with an explanation. This was Albert Einstein who, although he'd heard about Brownian motion, had never himself observed it. He had decided, nonetheless, that the source of the phenomenon was the movement of atoms.

If atoms really existed, Einstein thought, they'd betray themselves in the form of thermal energy. Every material body existed at a given temperature, which is to say it possessed a certain amount of heat. At the finest levels of matter, among the atoms, heat would show up as a kind of motion, a random thermal vibration. Although atoms and molecules (small groups of atoms bonded together) were individually too small to be seen, their motions were nevertheless strong enough to cause much bigger particles suspended in them to jostle around visibly.

It was like a bunch of people playing with an extremely large beach ball, all of them trying to push it in various directions — an analogy given by Richard Feynman in his *Lectures on Physics.* "We cannot see the people because we imagine that we are too far away," he said, "but we can see the ball, and we notice that it moves around rather irregularly."

In the same way, a moving particle in a stationary liquid was symptomatic of the thermal energy of the atoms surrounding it. Brownian motion, correctly understood, was visible evidence that atoms were real.

That, anyway, was Einstein's theory. But until there was some experimental proof for it, a theory was all it would ever be.

Einstein, however, proposed an experiment. If his explanation was correct, he said, then the particles suspended in the liquid would be spread out in a specific telltale fashion: there would be more of them at the bottom of the container than at the top. Gravity, he reasoned, would pull the particles down against the action of the moving atoms which tended to push them back up. All of this would take place in accordance with certain set mathematical equations — Stokes's law of fluid resistance, for one — and Einstein went so far as to specify, on the basis of those relationships, the exact numerical distribution of particles, varying according to height, that would be found in a given column of liquid.

Which was a rather amazing prediction. It was as if someone watching the motions of Feynman's beach ball had ventured to estimate how many people were batting it back and forth, and exactly where they were located on the beach — all this on the basis of an exhaustive study of the ball's motions. But it was not impossible if you knew physics.

Anyway, Einstein made his prediction and left it at that; though not precisely a bull in a china shop, he was not much one for lab work. Besides, this was the year in which he'd publish his special theory of relativity, state the formula $E = mc^2$, and explain the photoelectric effect — so he had better things to do with his time than stare at a column of water and tally up a bunch of particle counts. That little task he left to others: "It is to be hoped that some enquirer may succeed."

By 1908, someone had — a French experimental physicist by the name of Jean Perrin.

Perrin's work was made possible by a newly invented lab instrument called the slit ultramicroscope, which was a tool for investigating the fine structure of matter. By bouncing light off an object at right angles, and by passing the reflected light through a complicated series of slits, lenses, and beam-straighteners, the slit ultramicroscope could make even the tiniest particles (although not the atoms themselves) visible.

Perrin applied a slit ultramicroscope to the distribution of gamboge particles — bright orange blobs of tree sap — in water. He saw much the same sight that Robert Brown had seen some seventy-five years earlier, except that Perrin could now resolve the individual moving particles and photograph them, enabling him to count the blobs, one by one.

It was, you had to think, not one of the more glamorous jobs in science. You had this mental vision of Perrin showing up at the lab each morning, pulling on a white lab coat, and sitting down at the slit ultramicroscope for another fine day's work counting tiny dots.

But then again, much of science was plain grunt work, even the theoretical part. "Ninety-nine percent of theoretical physics is unbelievably dreary and grungy," said Hans Christian von Baeyer — himself a theorist. "Who wants to sit for a month sifting through pages of densely scribbled manuscripts looking for a lousy minus sign or a factor of two that has been dropped? Most of my experimental colleagues would rather sit over a microscope any day — at least that's reality, down there."

It was reality in front of Jean Perrin, at any rate. His particle counts confirmed Einstein's predictions to the letter. For example, when there were 116 particles at one level, there were 146 of them at the next lower level, then 170, and finally 200 at the bottom. It turned out that if the counts had only been 119, 142, 169, and 201, then they would have formed the exact geometrical progression forecast by Einstein. There was no way to understand these results other than by supposing that atoms were in fact real things, and that their random thermal vibrations were what kept the particles diffused through the liquid in that fashion.

"One might think," said science historian Steve Brush, "that this achievement would have made Jean Perrin an immortal hero of science, celebrated in all texts, encyclopedias and histories of science as the man who finally proved that atoms really exist. No such luck! There is a curious blind spot in our attitude toward the significance of various discoveries in science. Those who are fortunate enough to attach their names to quantitative 'laws' or 'fundamental constants' are assured of a place in the language of posterity. But a scientist who establishes a qualitative fact or concept — without which those laws and constants would be meaningless — is likely to have his contribution forgotten."

Perrin, although he'd win the 1926 Nobel Prize for physics, was indeed gone and forgotten by the end of the century, and few in the new generation of physicists had even so much as heard of him.

Some years after the existence of atoms had been accepted as fact, Albert Einstein finally saw Brownian motion for himself.

"It seems contrary to all previous experience," he said. "Examination of the position of one suspended particle, say every thirty seconds, reveals the fantastic form of its path. The amazing thing is the apparently eternal character of the motion. A swinging pendulum placed in water soon comes to rest if not impelled by some external force. The existence of a never diminishing motion seems contrary to all previous experience. This difficulty was splendidly clarified by the kinetic theory of matter" — the theory that matter consisted of atoms in motion.

But even after their existence had been proved through observation, atoms remained obscure and hidden. For one thing, atoms were so tiny — incredibly tiny — so very far removed from human-size scales as to vanish into something bordering on nothingness. To get a mental picture of how small they were was no simple task.

The "small" objects of everyday life were wholly colossal from an atom's point of view. Some of the smallest objects easily visible to the naked eye, crystals of table salt or grains of beach sand, for example, were fully half a millimeter (500 microns) or so across, and each one of them consisted of uncountable millions of atoms. Pollen grains, on the other hand, were roughly a tenth the size of a salt crystal — ten pollen grains laid end to end would give you one grain of salt from the saltshaker — and they ranged in size anywhere from 25 to 50 microns across. That was nearing the lower limit of human vision, 25 microns. Small as pollen grains were, though, they did not even begin to approach the truly impalpable minuteness of atoms.

Pollen grains, first of all, were not featureless pieces of matter like sand or salt. Instead they were complex entities containing within themselves lots of even tinier structures, organelles such as concentrated mitochondria, endoplasmic reticulum, the golgi bodies, and so on, plus the usual complement of genetic machinery normally found in living cells. Some of the smallest of these organelles were the ribosomes, whose function was to make new protein. Whereas the size of a cell was normally stated in microns (millionths of a meter), the size of a ribosome was normally given in the far smaller units of nanometers (billionths of a meter). An average ribosome was only about 25 *nanometers* across, a size that was so much smaller than the cell itself that in the typical schematic diagram of a cell, the ribosomes showed up as mere dots.

Small as the ribosomes were, each of them consisted of hundreds of thousands of atoms. Pollen grains, therefore, although they were almost invisible to the naked eye, were not "small." And even ribosomes, hundreds of times smaller than pollen grains, were not "small."

Atoms were small.

So small, in fact, that for the longest time there was no hope of ever seeing one — not by visible light, at any rate. Ordinary light, theoretically, would whiz past an atom just as if it weren't there. The reason was that light propagated out in waves, electromagnetic ripples that had a certain definite wavelength, an actual physical size, in the neighborhood of 500 nanometers from crest to crest. In order for an object to show up in visible light, it had to be large enough in cross section so that it would catch some optically meaningful portion of that light and reflect it back to the observer. But atoms were only about a tenth of a nanometer across — some five thousand times smaller than the wavelength of light. A single burst of light, in other words, was equal in size to five thousand atoms lined up in a row, so how could a light ray separate out just one of them?

Expecting to see an atom with visible light, then, was akin to standing on the edge of a canyon, clapping hands, and expecting to get an echo back not from the canyon wall opposite but from a withered twig on a distant tree. Atoms were invisible to light, not because of any failure of technology but because light itself was too gross for the task.

The other reason that atoms were foreign to human beings was the fact that they were in constant motion. There was no such thing as rest or stability down there, and there was no good analogue to Brownian motion up here, in the Big World. Atoms could be known, it seemed, only through physics.

Modern physics had a shorthand expression for the kinetic energy of a single atom at room temperature: kT, where k was a constant (Boltzmann's constant) and T the temperature in degrees Kelvin. The higher the temperature, the faster the atoms would vibrate back and forth, exhibiting their jittery little *kT* energy levels. It didn't take any especially high temperatures in order to get some rather goodly rates of vibration going, and the physicists who studied Brownian motion — Einstein, Smoluchowski, and others — calculated that at room temperature a single molecule of water — one molecule — experienced roughly 10^{15} collisions per second with other, surrounding molecules.

By any standard, that was a lot of collisions. Indeed, at room temperature, life at the molecular level was a sea of assault and battery, and what appeared, at the macrolevel, to be utter calm and stillness, was in reality the scene of the worst cosmic thrashing imaginable. Nobody would want to go through life as a molecule.

It was this precise fact, that life at the molecular level was hell, which explained why humans were as large as they were, at least in comparison to atoms. The physicist Erwin Schrödinger had asked the question back in the 1940s: "Why must our bodies be so large compared with the atom? Is there an intrinsic reason for it?"

The answer was yes. Human beings were as big as they were so that they'd be oblivious to the chaos that lay down below, in and among the moving molecules. Just try to imagine the opposite situation, he suggested. "What would life be like," Schrödinger asked, "if we were organisms so sensitive that a single atom, or even a few atoms, could make a perceptible impression on our senses?"

It would be like being battered senseless by 10^{15} collisions per second — a "funny and disorderly experience," as he put it. "An organism of that kind would most certainly not be capable of developing the kind of orderly thought which, after passing through a long sequence of earlier stages, ultimately results in forming, among many other ideas, the idea of an atom."

The other side of the coin, though, was that insensitivity to atomic jostlings meant an equal unfamiliarity with the whole atomic realm. It was an alien domain, terra incognita. You had no intuitive concept of it.

Which was just as true for chemists and physicists as it was for anyone else. Scientists tended to regard atoms as abstractions — as the stuff of equations, formulas, and symbols like *kT* — instead of as real and physical objects. They could quote you chapter and verse of the chemical bonding rules and laws of valence. They could recite all sorts of facts and figures about atomic weights and numbers, the rules of beta decay, and so on, but as to how a given atom behaved *mechanically* — as a thing that you could physically touch, push around, and manipulate — well, here they were as much at a loss as the next person.

You'd ask an MIT chemistry professor — a well-known chemist, tops in his field — you'd ask him: "If you had a carbon atom, just a typical carbon atom lying around on some surface — not attached to that surface in any kind of bond, which probably isn't possible, but if it *was* — then how much would it vibrate along its own length at room temperature? Would it vibrate back and forth an angstrom, for example?"

And the professor's answer was: "I have no idea."

No idea?

So you'd ask him: "But is this a known quantity? Would somebody be able to tell me?"

And he'd say: "You're not talking to the right person. These sorts of questions . . . it's like talking to an engineer about the fundamental laws of physics. They're not something that an engineer — say a civil engineer or a bridge builder — would know. You learn them in school, but you don't use them every day."

The modern chemist, with rare exception, dealt with extremely large statistical quantities of molecules: test tubes, flasks, beakers, comparative oceans full of molecules, never with just *one*.

And from all this — from the tininess of atoms, from the fact that they were in constant motion, from the fact that you never dealt with just one, or two, or any discrete number of them — it seemed to follow beyond a shadow of a doubt that nobody in his or her right mind could possibly think of "working" with atoms, or of "building things" with them, or of doing any "engineering" with individual atoms. Any such person was mildly insane, and in no case could he or she be thought to know the first thing about the greater glories and mysteries of science.

That, then, was the situation confronting Eric Drexler at the very moment that he was first coming up with his grand idea, the idea of a molecular technology, which involved working with atoms one by one and manipulating them exactly as if they were the literal "building blocks" of all things, not just the metaphorical ones. What was a scientist to do when hearing about such a scheme other than relegate it to the comic strips?

The irony of it all was the way in which *kT*, thermal energy, had turned out to be nanotechnology's friend and nemesis, both blessing and curse. If it hadn't been for *kT*'s showing up in the form of Brownian motion, humans might never have gotten a clear proof that atoms existed, in which case, obviously, the idea of nanotechnology would never have arisen. But on the other hand, it was that very same *kT* — those incessant molecular pummelings at rates of 10^{15} per second — which made nanotechnology seem, in the minds of working scientists, to be an entirely deranged proposition.

Life was hell down in the molecules, that was for sure, and a nanodevice would be like a leaf in a hurricane. However pretty a picture it made, nanotechnology was still only "science fiction."

Not that this ever daunted the committed nanotechnologist.

"Oh, 'science fiction,' is it?" said Ralph Merkle. "Is that like, you know, *Going to the moon?*"

2

The Old Technology

When he first showed up at MIT, K. Eric Drexler was still firmly in the grip of the "old" technology — chemical rocketry, megabuck space stations, heavy-lift launch vehicles blasting across the solar system. Not that you could blame him: this was the fall of 1973 and the heyday of manned space exploration, the year in which American Skylab missions were keeping astronauts in orbit for months at a time. But however "futuristic" it had once sounded, by the mid-1970s space travel was pretty much old hat. As for chemical rockets, they were very novel items back in the days of ancient China. There was nothing essentially new about any of this.

To Drexler, though, all of it fit in squarely with his own personal vision, his dreams of mankind going out and subduing the cosmos. MIT was the perfect home for him and his dream. The place suited him to a T, it was a strictly business, one-dimensional science-and-technology school. On the institute's buildings, in the so-called entablature, where names like Plato, Shakespeare, and Beethoven would go if this were a liberal-arts school, there were, instead, names like Archimedes and Darwin, Newton and Faraday, Pasteur and Lavoisier. Those names, chiseled in two-foot-high block lettering, pretty much gave you the tenor of the place.

It rather lacked an aesthetic dimension: there were no tidy red-brick buildings knee-deep in fallen oak leaves, no quaint cupolas, no lush green groves bordered by tall pines, no horses gamboling across college pastures. MIT had none of that. Physically, it was a triumph of stone and concrete. Every other building looked like a mausoleum or an industrial plant, or like it housed the all-government Bureau of Ink, or some such thing.

Then, too, MIT was the home of the nerds — it was internationally famous for this. It got to be so bad that at one point the institute took matters into its own hands and offered a Charm School for its worst offenders. That was what they called it: *Charm School,* sponsored by the Office of Undergraduate Academic Affairs. According to the ads that ran in course timetables, this unique, cost-free, walk-in service would teach you such things as Facial Affect, Gentle Self-Deprecating Irony, Cheap Tasteful Gifts, Asking for a Date, and the crucial technology of Salutations — beyond "Uh, hi" and "How ya doin'?"

Not that Drexler himself was especially nerdy: he never once in his life wore a white short-sleeve shirt and plastic pocket-protector stuffed with two hundred ballpoints. Not once did he ever look like that. He was rather in the sixties mode: black T-shirt, jeans, beard and ponytail. From outward appearance he was a poet or artist, and in fact he did have some artistic tendencies. He was particularly good at sculpting, and once produced a clay figure of Richard Nixon's head that was also fully functional. You blew through a mouthpiece at the back and out of the head came the sound of a train whistle. He was not lacking a sense of humor.

Internally, though, he was the soul of engineering and applied science. He wanted to learn science not for its own abstract, intellectual sake, but rather for what you could physically do with it. And what Drexler more than anything else wanted to do with science and technology was to get the great space migration going, launching people off the planet and out into the solar system, where they could flourish in peace, freedom, and material comfort.

Anyway, when Drexler first arrived at MIT he took up residence in Senior House, one of the undergrad dorms at the east end of campus. This was a large, stone, L-shaped structure divided into six different sections. He got a room in the so-called "Runkle" section, room 106 on the ground floor. He had no roommate.

One of the first things you did as an incoming freshman was to see your adviser. So one day Drexler went in and saw his adviser and told him of his deep and abiding interest in space travel, space colonization, and rocketry. The adviser told him to see Philip Morrison, the physicist, who was rather like-minded.

That turned out not to be true. Morrison was known for two things. He was known as the book reviewer for *Scientific American*, and he was known for his role in the Manhattan Project. He had once held, in his very hands, the plutonium core for the world's first atomic bomb: it had been warm to the touch, a consequence of radiation. Later, he'd sat next to the plutonium core — it was enclosed in a small, shockproof box — as it traveled in an Army sedan from the Los Alamos lab, where it had been constructed, to Trinity Site, where it would be detonated.

One of the things that Morrison had acquired a great sensitivity to, as a result of these experiences, was radiation. There was a lot of radiation out there in space, however, and so Morrison was not what could be called a great big fan of space settlement. But he knew someone who was: Gerard K. O'Neill.

O'Neill was in fact the world's number-one proponent of space colonies. He wanted to build these enormous, twenty-mile-long artificial habitats up in earth orbit. Others had had such ideas before, of course, but O'Neill was the first one to flesh out the concept scientifically, with facts, figures, formulas, and even some preliminary engineering designs.

Drexler was highly interested. The only problem was, O'Neill was not at MIT. He was at Princeton, where he was professor of physics.

So Eric Drexler went down there to see him. The nineteen-year-old Drexler and the forty-seven-year-old O'Neill got along famously right from the start, and by the end of his freshman year Drexler was presenting his first scientific paper, "Space Colony Supply from Asteroidal Materials," at a space settlement conference that O'Neill had organized.

"I can picture dinner at the O'Neills'," Drexler recalled much later. "Freeman Dyson was there. We were served salad from a stainless-steel bowl that was supposed to have been part of a particle accelerator that O'Neill had been working on. But they'd machined too many of these things, and so the O'Neills were using them as salad bowls."

Soon Eric was leading this complex, three-cornered life that took him from MIT, where he was a student, to Princeton, which was his spiritual home, to Mountain View, California, which was the home of Gerry O'Neill's summer study programs — ten-week-long marathons of space colony design theory. For a while there was one of these every summer, held at the NASA Ames Research Center, on Moffett Field, south of San Francisco.

It was a naval air station and so the place gave you something of the feeling, the mood, the ambience, of space travel. There was this big wind tunnel at one edge of the base, a truly gargantuan structure that looked properly otherworldly, as if an extraterrestrial landing craft lay hidden under wraps. Sitting in the seminar room with O'Neill and the other space cadets, and with jet-engine whine for background music, you could actually imagine heading off to the stars.

Back at MIT, Drexler emerged as the institute's chief space leader. He was not what could be called a charismatic personality — not exactly — but he seemed to have a born talent for communicating a vision. And because he was so fabulously successful at this, at winning people over with his knowledge and ideas, he tended to attract a following. During the January breaks, he taught informal MIT courses on space colonization — this as a part of the school's Independent Activities Period, which gave students the opportunity to do such things. Starting with his freshman year, Drexler always availed himself of the chance, teaching an IAP course every winter. His space colony offering was so popular that he'd soon founded his own space society, the MIT Space Habitat Study Group.

Kevin Nelson was one of the founding members. "Eric ran the first meeting," Nelson recalled. "He started off with some introductory remarks, then asked people to say what their interest in the subject was, and then he went around the room. The first person said, 'I've always been interested in space and spacecraft.' The next one said, 'I design spacecraft for a living.' The third one said, 'I fly spacecraft for a living.' "

That was semi–tongue in cheek: the speaker turned out to be Phil Chapman, a former NASA astronaut who was then working in Boston. But that was just the type of specimen you had at your disposal at MIT: the gung ho, dedicated, educated techie.

For a while, at these Space Habitat Study Group meetings, all you did was sit and talk. But *then* . . .

"At one of the meetings," Nelson recalled, "we sat down and we asked, 'Why aren't we making more progress? Is there something else we can do that would be more effective?' I made a grid of all the possible ways, unlikely as they were, that we could put up a space station. We could convince the military that it was needed, we could convince Congress that it was needed, we could convince private industries that it was needed, we could convince stockholders and private companies, we could convince voters to vote for Congressmen.

"Or we could do it ourselves."

MIT Kids Launch Space Station!

"We sort of snickered about that last one," he added. "But as we thought about it we realized that Max Faget, NASA's chief spacecraft designer, was just then getting ready to leave NASA. We knew astronauts, we knew engineers, we knew Wall Street people, and we knew think-tank vice presidents. And so we decided that it wouldn't be all that ridiculous to get these people together and sit them down in one room and see what they thought of the idea, whether it was practical."

Which it was not. But then again, neither were NASA spacecraft "practical." It took countless billions of dollars to get a couple of astronauts up in orbit. Even the vaunted space shuttle, NASA's bargain-basement "space truck," could not be called especially thrifty. On the rare occasions when it actually flew, it took the astronomical sum of $10,000 per pound to place a given payload in space. That was not "practical."

But then again, it was only what you could expect with the *old* technology.

* * *

The primary entities of the new order of things, meanwhile, had been gradually emerging from out of the gloom. After the atoms had been proved to exist, the first thing scientists wanted to do was, of course, to see them. While everyone was convinced that this was impossible with light, that was only true of *visible* light. But visible light was just one type of light: there were also others — ultraviolet light, for example. Visible light, in fact, occupied only a tiny portion of the broader electromagnetic spectrum, which, in addition to ultraviolet, included gamma rays, X rays, infrared radiation, and radio waves. There was an entire range of electromagnetic radiation available, and conceivably some of it could be utilized to picture the atoms.

The first person to put this idea to the test was a German physicist by the name of Max von Laue, a man who prided himself on his knowledge of optics. "As far back as I can recall having had an interest in physics," he once related, "my particular attention was drawn to the field of optics, and within that field to the wave theory of light."

Visible light was no good for seeing the atoms because its wavelengths were too big. To picture atoms you'd need to use radiation of a much shorter wavelength, one that was about the size of the atoms themselves. It was a case of fitting the spotlight to the quarry.

There was, fortunately, a type of electromagnetic radiation whose wavelength was almost identical in size to an atom: the X ray. The average atom was about an angstrom across — a tenth of a nanometer — and so was the wavelength of the average X ray. Sizewise, atoms and X rays were perfectly matched.

In 1912, Max von Laue had the idea of beaming X rays through crystals, thereby revealing the atoms in all their glory. Accordingly, he took a crystal of copper sulfate and irradiated it with X rays. An image appeared, sure enough. The question was, what was it an image *of?*

Photographing atoms with X rays bore little resemblance to the common practice of taking an X ray picture of the human body, where the results were clear and intelligible even if spooky. An X ray picture of atoms in a crystal lattice — the atoms arrayed out geometrically in columns and rows — bore only the most indirect and tenuous relationship to their actual physical layout within the crystal. This was due, among other things, to the phenomenon of wave interference, a consequence of the X rays reflected back from one layer of atoms being out of phase with those reflected back from the next layer below. The net results were some weird and ghostly patterns, mere shadow-skeletons of the underlying atomic arrangements.

Von Laue never got any really good pictures of atoms, but at least he confirmed the fact that the wee creatures were actually down there. If they weren't, no interference patterns would have appeared; the X rays would have traveled straight on through.

One thing Drexler did on a regular basis as an MIT undergrad was to pay nighttime visits to the science library.

"Like any sensible person," he recalled, "you obviously want to understand more about your field and so you read the things that come through the science library every day, trying to find things you don't understand, and you read enough about them so that you do understand. This is, I'm sure, what scientists and engineers do, because otherwise you don't understand science and engineering."

Later in life he'd have reason to draw back a bit from that rashly optimistic conclusion, at least as it applied to one or two scientists whose names he could mention. But *he*, at any rate, always wanted to keep up with things, so late at night he'd take himself over to the Hayden Memorial Library, a squarish, modern-looking building a short lope from Senior House. He'd plant himself in the periodicals room, an area so bland and boring that the only conceivable activity there was reading, and start leafing through the various journals. There were lots of them on the shelves, journals like *Bioscience, Progress of Theoretical Physics, Journal of Organometallic Chemistry,* and so on. There were, in addition, more popular items like *Science, Nature, Scientific Amer-*

ican, Physics Today, and *Science News.* Drexler would page through these, one by one, to find out what was going on in science.

What was going on, at the time, was the revolution that came to be known as genetic engineering. This was the attempt to alter an organism's genes so that they'd produce something different from what they would normally. The genes in bacteria, for example, might be modified or replaced by others so that the organism produced human insulin instead of yet more bacteria. However strange and "against nature" it may have sounded, such processes occurred all the time in nature — as, for example, when a virus reproduced itself. Viruses, because they lacked the necessary metabolic equipment, replicated themselves only by commandeering the biochemistry of a host cell. The virus injected its own DNA (or RNA) into the cell, causing the cell to manufacture copies of the virus instead of further copies of itself. That seemed "against nature," too, when you thought about it, but it occurred in nature all the time. Genetic engineering was merely the attempt to perform similar feats by design and intention.

Since the genes were contained in the DNA molecule, modifying a gene meant modifying the DNA. Although it was the molecular basis of life, DNA was not itself a living thing; rather it was just a long and complicated chemical structure, an extremely large molecule. It consisted of about a hundred billion atoms strung together in two long strands that twisted around each other — the famous double helix.

The important thing about DNA, however, was what those long strings of atoms represented — namely, information. The information stored in DNA was the software, the programming, that ran the body's cells, from which it seemed to follow that if you could rewrite the programming, then you could conscript the cell machinery and get it to do your bidding, just as the viruses did. All you'd have to do was to change the sequence of the atoms that made up a given molecule of DNA.

By the early 1970s, scientists had learned how to do precisely that. The key to it all was the discovery of two families of enzymes — biochemical catalysts — that performed extremely specific and highly controllable functions. One type, the so-called restriction enzymes, had the ability to slice through both strands of a DNA molecule at a predefined location, physically splitting it up crosswise into two separate fragments.

The other enzymes were the ligases. These had the exact opposite ability: they were able to join together any two DNA fragments, thereby forming a larger molecule. Those two chemical substances — restriction enzymes and ligases — were the primary tools that scientists needed in order to break up a given DNA molecule and put the fragments back together again, all according to plan.

The most amazing part of the process, though, was that the fragments to be joined together did not have to originate from one and the same organism. They did not even have to originate from the same *species* of organism. In fact, since DNA was just a big molecule, just a long chain of bonded atoms, the pieces to be connected up did not even have to come from an organism at all: it was chemically possible to synthesize a single short strand of "artificial" DNA and then to add it to a strand of a "natural" DNA, giving you a completely new molecule.

Soon enough, all of this had actually been done in the lab. The DNA from one species of bacteria had been joined to the DNA from that of a different species. The DNA from a monkey virus had been joined to the DNA from a lambda phage. And the genes from bacteria had been joined to those of fruit flies and frogs.

As it happened, all of these strange and wondrous feats had been performed in 1973 — the very same year Drexler entered MIT.

So on his nighttime visits to the science library, Drexler acquired a working knowledge of genetic-engineering techniques. It was, he discovered, a complex business with its own highly refined vocabulary, a language of plasmids and EcoRI enzymes, shuttle vectors and reverse transcriptase. But he was quite suited to learn it all, being an "interdisciplinary science" major and well versed in many branches of science. Still, it took him a couple of years.

By the time he got to the end of it, though, he'd gotten some ideas of his own about what could be done with these techniques. The fact was that for all of its potential revolutionary implications, the point and purpose of genetic engineering was really quite modest. Basically, genetic engineers wanted to direct their techniques toward the making of biological substances that were useful to mankind. They wanted to reprogram bacteria to produce things like insulin, human growth hormone, or vaccines.

Entirely laudable goals, of course, but still and all pretty humdrum. Anyone could imagine producing human growth hormone; there was nothing too electrifying about the idea. But by the end of 1976, by which time he'd learned enough about genetic-engineering techniques to see their true potential, Drexler had his own private notion as to what *could* be made with some intelligently reprogrammed DNA.

Not insulin. Not hormones. But rather . . . a computer.

When the first genuine images of atoms appeared, few could have known the form they'd take. It would not have seemed a reasonable proposition, before it happened, that one day you'd be able to sit in front of a cathode-ray tube — basically just an ordinary television screen — and behold atoms in real time. Not broadcast images of atoms, but images of the atoms right in front of you, the very ones of which the cathode-ray tube itself was composed. You'd watch the screen and see with your own eyes the atomic structure of the hot metal cathode below.

Improbable as it seemed, by 1955 it had actually been done, by Erwin W. Müller, a physicist at Penn State University.

Müller had gotten his doctorate in physics from the Technical University of Berlin in 1936, at the age of twenty-five. He'd done his work under physicist Gustav Hertz, who steered him toward an area known as "field emission," which was the process by which electrons were emitted from the surface of a conductor in the presence of a strong electric field. Shortly thereafter, while working for Siemens, the German electronics firm, Müller got the notion that he'd like to see this process for himself, as and when it was actually occurring. He decided that if you applied a high enough voltage to the needle tip of a cathode-ray tube, then the tip's surface electrons would fly up toward the fluorescent screen, where they'd leave an image of some sort. Since the electrons would travel in relatively straight lines from the needle tip to the screen, the image produced ought to bear some visual relationship to the needle tip itself.

Which it did. What showed up after he'd finally gotten the instrument to work properly (it took some years) was a highly magnified picture of the needle tip's surface. Müller called the device a field-emission microscope.

The field-emission microscope was the most powerful microscope in existence, but it was a microscope of a new and unusual sort. For one thing, the only thing you could see with it was part of the very device itself. And instead of operating like a normal light microscope, where you shined light on a specimen and viewed it through a lens, this one worked quite differently. Portions of the specimen itself were sent flying toward the screen, upon which they left their fleeting traces. But then again, an ordinary light microscope would never let you see what Müller saw.

Müller's device resolved features that were as small as 20 angstroms across — roughly the size of twenty atoms lined up in a row, the scale of medium-size molecules of matter. As he worked with his device, Müller saw not only molecules, he also observed their heat motion, which is to say, *kT.*

"We can actually see the thermal agitation of the molecules," he reported. "By gradually varying the temperature of the needle, we can watch the changes in molecular behavior."

But he could not quite make out individual atoms. That was hard to do with the field-emission microscope because of the fact that the electrons coming up off the needle tip interfered with one another's motion slightly. They didn't travel in perfectly straight lines, and so they created a somewhat blurry picture. To see the atoms, he decided, you'd have to use particles that were far heavier than electrons. So Müller tried using ions. Ions were atomic nuclei, roughly two thousand times heavier.

This new and improved device was the "field-ion microscope," and by 1955 it was able to produce pictures of atoms for the first time in recorded history.

"The atoms constituting the specimen surface can be seen individually," he said, "a feat not possible with any other type of microscope presently known."

The atoms showed up as white dots on the screen, arranged in various regular patterns. A photograph taken with Müller's field-ion microscope ran on the cover of *Scientific American* in 1957, with the caption "Atoms visualized."

The whole thing was, in retrospect, a marvel of simplicity. No multimillion-dollar cyclotrons were required, no extremely complicated equipment. In fact the entire setup was improbably crude. Basically you needed a cathode-ray tube, a finely controllable power source, and a darkened room. ("The best way to observe the screen is directly with our eyes, after they have been conditioned by a long period of darkness," said Müller.) And with those things you could see the atoms.

It was a pattern to be repeated again and again in the quest for ultimate control over the constituents of matter. An objective that was, from the looks of it, exceptionally difficult if not absolutely impossible of attainment, yielded itself up to ordinary instruments and easily reproducible techniques. Atoms might have been bashful at first, but they submitted themselves to the inquirer upon being approached correctly.

Biology, Drexler thought, could somehow be cajoled into producing a computer.

Why a computer, of all things?

"Well, computers for many years had been getting better by getting smaller," he explained much later. "Miniaturization is a big theme in technology, so this is a fairly obvious thing to think about. I mean if you're thinking about making small things, the first thing that comes to mind is a computer: computers are supposed to be small. People were spending zillions of dollars trying to make small computers so it's apparently a good idea to make small computers. Other things being equal, smaller computers are faster and more efficient."

All of which was entirely true: other things being equal, there was no particular virtue in largeness. The bigger an object was, the more space it took up, the more resources it consumed, the more power it took to operate, the more waste heat it generated, the slower it worked, and so on and so forth. For the specific purpose of getting the same bit of information from here to there, the hummingbird was more efficient than the eagle, and the bee than the hummingbird. The same principles held true for computers, whose sole purpose in life was merely to manipulate various bits of information. A bit of information had no inherent size, and so the smaller you made the apparatus, the better off you were.

"Now the smallest components in the world are molecules," said Drexler, "and if you look for complex, organized systems of molecules you find them in biology. And if you study what the principles are, you conclude that you can make very different — and in some ways more highly organized arrangements of those molecules — and it's natural to think of using those arrangements to make computers."

At least it was natural for him. But in truth, after you'd considered the idea for a while, there was a way in which the notion of getting DNA to make computers was in fact natural in the extreme. What was an animal's brain, after all, but a biological computer, one that had been manufactured by the DNA and its associated cellular mechanisms? Drexler's molecular computers would be different from brains in that they'd be mechanical as opposed to biological, but they'd also be tinier, faster, more efficient, and have all the other advantages that a smaller machine had over the relatively massive human brain, which, after all, weighed three pounds, was slow and unreliable, and lost cells on a daily basis, never to be replaced.

Drexler's key insight was not so much his regarding the components of biological cells as machinery — by now that was commonplace — but rather his regarding them as tools with which you could get the cells to manufacture a completely new type of thing, as opposed to just a slightly different type of old thing.

"The idea of molecular machinery was very much in the air," he said, "except that the people who were studying it weren't engineers and so they weren't thinking about building things with these devices. But these entities — protein structures, enzyme reaction mechanisms, and so on — they were clearly devices, mechanical and electronic widgets inside the cells. They were synthesized chemically by cells, they spontaneously assembled inside cells — or even in the test tube: you mix the parts together and through selective stickiness they stick together to make complicated little devices. They were clearly physical objects that could go together to build larger structures.

"So I started asking myself, 'Well, what if *we* could do things like that?' I asked myself what these biological components could do if you were able to design them, and get them to do, locally, the kinds of things you see them doing in nature, but yet have those parts fit together in a new way to make a very different pattern globally. Not a pattern of living things but a pattern of regular structures and devices doing things like, for example, computation.

"And at that time — this would have been in late 'seventy-six — I thought that you could self-assemble multiple layers of biomolecule-type things, and make a very small computer. I was persuaded that with a sufficient amount of effort one could develop computers on a scale of nanometers, with components based on the kinds of physical phenomena that are seen in biological materials, but without the structure itself being a biological system. They would be self-assembled objects made out of proteins, or proteinlike molecules with lipids. I didn't have any marvelous insights, I just had this image of self-assembling things."

That was how far he'd gotten roughly by the Christmas of 1976, at which point he was an MIT senior.

Christmas came and went, and then soon after the New Year Drexler got an even better idea. This new notion truly staggered even him, and he was not a man easily impressed.

A computer was a mere mechanism, just one particular type of machine. There was nothing inherently special about it: a computer was just a given arrangement of matter that worked in a certain way. But if you could induce biological mechanisms to manufacture one type of nonbiological entity, then you could probably persuade them to manufacture other types as well.

And if that was true, then where was there a stopping point?

Why couldn't you program biological mechanisms to produce anything you wanted?

DNA, certainly, produced lots of different things already. In fact, it had given rise to all the plants and animals of the entire animal kingdom. Life forms as different from each other as ants and whales, elks and daffodils, viruses and starfish — all of them ultimately stemmed from programs rolled up in molecules of DNA. So if by manipulating matter at the molecular level you could induce biology to build molecular computers, and if biology already built all kinds of viruses and bacteria and plants and fishes and animals . . . then why couldn't its already broad repertoire be broadened further still? Why couldn't biological machinery be programmed to make all sorts of nonbiological entities?

"By early 1977 I had realized that instead of information processing, the really interesting, high-leverage application of molecular machinery was to manipulate molecules to build other things, including better molecular machinery. If you iterate this process, of using machines to build better machines, it's pretty clear that you could build machines that could build copies of themselves.

"Ultimately you'd be able to get macroscopic quantities of mechanism out of that self-replication process, and so you'd be able to get macroscopic quantities of other products, too. So in a series of little realizations and bits of the puzzle fitting together — this was in the spring of 'seventy-seven — it became clear that you could have what I would now describe as a molecular manufacturing technology. You'd have a molecular technology that could be used to manufacture a wide range of products.

"That very rapidly began to look like something very big and important," he added. "Because if you have this very general ability to manipulate atoms in complex patterns . . . well, then you can make essentially anything that's physically possible."

Anything that's physically possible.

Even to K. Eric Drexler, that seemed a rather grand and sweeping statement. Somehow it seemed too good to be true.

But then again, what was wrong with it? Where was there a false assumption? An error in the logic? Nowhere he could see.

3 The World-Class Auto-da-fé

They were a little weird, Drexler's thoughts were, after he'd gotten his new idea. He'd be walking back from the science library — the Charles River on his right, the lights of Boston off in the distance, the winter wind in his face — and he'd be turning these matters over in his head. Even to him there was a slight bit of unreality surrounding the notion. It seemed, in a way, too big a concept, as if there were too much that could come out of it, that it was too pregnant with possibility to be entirely real.

Molecular computers were one thing, they sounded realistic enough. But a whole technology based on molecular machines? Machines with which you could make practically anything? How likely did *that* sound?

So he reviewed his reasoning again and again. Step one was the premise that genetic-engineering techniques could be improved to the point where some new and novel molecules — specifically, new proteins — could be synthesized by the cells. Since biotechnology was already such a rip-roaring success, that much of the argument seemed uncontroversial.

Step two was the supposition that if those molecules were properly designed, then under the influence of Brownian motion they could be made to assemble themselves into more complex structures — such as, for example, computers or even more complex programmable molecular devices. Since there were many well-known examples of self-assembling artifacts in the biological world, this point, too, seemed beyond objection. Molecular biologists, Drexler knew from his nighttime readings, had broken down ribosomes (tiny subcellular organelles) into more than fifty separate components and had mixed those components together again, whereupon the ribosomes spontaneously pieced themselves back together. It was all a matter of appropriate design, of the parts fitting together properly, lock-and-key fashion, and so on. But if biological artifacts could self-assemble all on their own, why couldn't intelligently constructed machinery be made to do likewise?

The third step of the argument was the assumption that those self-assembling programmable machines, once you'd created them, could be made to build even better, more complex, even more dexterous mechanical devices. Ultimately they could build machines that could replicate themselves, that could turn out any arbitrary number of identical copies. But that assumption, too, seemed utterly reasonable in light of biology: the course of evolution showed that extremely complex functional entities (like animals) could arise from much simpler things (like the primordial slime). As for self-replication, if there was anything that biological cells were good at, it was this.

And from those three premises taken together the ultimate conclusion seemed to follow inescapably that it was scientifically possible to create controllable molecular machines that would do pretty much whatever you wanted them to, machines that could be programmed to manufacture anything allowed by natural law.

"Very early on," he recalled, "it started to look like a conclusion that was so solid that it was very hard for me to see how anybody could argue against it. There just wasn't anything new there in a scientific or physical sense, but only in an engineering sense."

And so at length the suspicion grew in his mind that he'd made some major new sort of discovery, one that he'd come up with all by himself, one that nobody else, apparently, had even the least inkling of — a fact which rather surprised him.

What you did in circumstances like that, of course, was to tell all your friends and neighbors about this tremendous new insight of yours, which is precisely what Drexler did. He told his old friend from high school, Dave Anderson, who every so often came up to visit him from New York, where he was an economics major at Columbia.

"Right from the start he was very excited," Anderson recalled. "He would tend to gesture — he's kind of a hand-waving guy a lot of times when he's trying to describe things, and I remember him gesturing a great deal and drawing pictures, and even the tone of voice. I know when Eric's excited about something by his tone of voice. He may sound a bit monotone, but I can tell the difference. Anyway, he talked at great length about it and he seemed to be full of ideas. It wasn't just a general picture he had, but he seemed to be full of all sorts of detailed thoughts about how this was going to work."

He told his friends in the Space Habitat Study Group.

"He would talk about molecules, their behavior, what you might get them to do, how you could control them," said Kevin Nelson. "He'd say, 'Well, molecules behave in this way,' as if he were a teacher. 'Imagine if you could make them do this,' and so on. He'd want to see if you were following the same line of thought as he was. He wanted to see if you jumped to the same conclusions. He was sort of laying out a trail of bread crumbs to see if it looked like a trail to other people."

"He talked generally about the idea of molecular manufacturing," said Dave Lindbergh, another study group member, "of building things up molecule by molecule, of the idea of an assembler — although I don't know whether he actually called it an assembler at that time. He talked about lots of neat applications, too, and my reaction was, 'This is really a neat idea.' "

And he also told his girlfriend, whom he'd met a couple of years before. He'd been walking down the second-floor hallway of his dorm, looking for someone from whom he could ask a favor, when he beheld, coming around the next corner, a most welcome and unexpected sight at MIT: a young woman.

He introduced himself and struck up a conversation. Her name was Chris Peterson, she said, and she lived in room 201.

"I wonder if I could ask you a favor," said Eric. "There's this neighbor of mine who's an Orthodox Jew and I'm supposed to light the stove for him for the Saturday-night dinner, but I have to go out of town that night, so could you . . . ?"

Yes, she said, she'd light the stove, no problem.

They met up again when he got back from his trip — he'd gone down to Princeton — and they started hanging around together. She was a freshman, eighteen, slim, and had long, brown hair; he was a junior, twenty, slim, and had long, brown hair.

He talked to her about graphite.

"Graphite!" Chris Peterson said. "He told me about graphite interminably! He was into space systems, he was designing space systems, and evidently at some point he became interested in graphite as a building material. He would go on at great length about this and I thought . . . Well, I mean, I was not all that interested in graphite, but he was really excited about it. Ask him about graphite and he'll tell you in more detail than you want to hear about it, believe me."

At the end of the school year the two of them went their separate ways for the summer — she back home to Buffalo, he to Mountain View for Gerry O'Neill's summer study. But they stayed in touch by mail.

"He would send me these postcards," Chris Peterson said. "Or maybe letters, I really don't remember, but I would periodically get these missives from California. Which was a good thing, because at that time we weren't a couple yet and so it was good that I was reminded, 'Oh, yeah, Eric Drexler. Graphite.' "

But when Chris and Eric got back together again the following year, Eric wasn't talking anymore about graphite. Now it was all molecular machines.

"I wasn't paying close attention to this," she recalled later. "I didn't know it was that important. I mean, what are the chances? Here is some senior in college talking about his ideas for molecular machinery, and I had no reason to see why they were important at that point. They were just his ideas and I didn't see the difference between lightsails and lunar processing and this new stuff."

* * *

Eric graduated in 1977 with a bachelor's degree in interdisciplinary science, and then started in immediately on his master's program. Grades had never been his strong suit — he spent too much time in college libraries for him to make straight A's in courses — but they were good enough for him to get a National Science Foundation fellowship, the first one ever granted in space industrialization.

For the time being, Eric decided that he'd remain in conventional mainstream technology — the "old" technology — rather than try to pioneer these new molecular-engineering ideas of his. The fact was, he'd gotten some *other* new ideas, these within the field of aerospace technology itself. One was for a new type of propulsion device, the "lightsail."

One of the great drawbacks of standard chemical rocketry was the small size of the final payload as compared to the total mass of the fuel required to get it off the earth and up into space. The three-stage Saturn V rocket that had put men on the moon, for example, was pretty much all fuel. The first stage weighed in at some five million pounds, was almost 100 percent fuel, and all it did, essentially, was boost the remaining two stages off the earth's surface: it burned for less than three minutes and then dropped like a stone into the Atlantic. The second stage weighed about one million pounds, was also almost 100 percent fuel, and all *it* did was to boost the third stage to the point where it would cut in and take over. The final payload — the command, service, and lunar excursion modules — had a combined weight of only 95,000 pounds, which was almost exactly equivalent to the amount of fuel the first stage consumed during its first 8.9 seconds of engine burn, before it even left the launch pad. Traditional rockets, in other words, were some of the biggest wasters of chemicals in history.

But Drexler had found a way around that, at least for travel between two points in free space (getting from earth to orbit was another matter). He wanted to use sails for propulsion, just precisely the way sailboats did, except that instead of wind providing the push, sunlight would. You'd attach the payload to a large, thin reflecting surface, then sail around the solar system on a steady flow of sunshine.

The idea was by no means original with Drexler. It went back to the Russian space theorists Tsiolkovsky and Tsander, who'd proposed using solar sail propulsion back in the 1920s. Solar sails had never actually been used, though, the main reason being the stupendous dimensions of the sail required. The pressure of sunlight falling on a surface was so weak that an effective solar sail would have to be rather sizable to move even the most modest cargoes: a surface the size of a football field was required to accelerate a payload as small as a marble, for example, and a sail roughly a square kilometer in area would be needed to drive a one-ton payload through space. As Freeman Dyson had once observed, "Nobody wants to be the first astronaut to get tangled up in a square kilometer of sail."

Despite that unnerving prospect, NASA had given serious consideration to solar sails, coming up with a scheme that involved building the sails on earth, folding them up, launching them into space on chemical rockets, and unfolding them again in orbit. The problem with this was that in order for the sail to survive the successive rigors of folding, launch, unpacking, and deployment, the material itself, a thin foil, would have to be beefed up considerably with a plastic backing. This would make the sail bulky and heavy — which in turn meant lots of chemical fuels, waste, and expense, somewhat defeating the whole purpose.

Drexler's idea was to keep the sail as light as possible. He envisioned an aluminum foil so very thin that it would be only a few nanometers thick — as little as fifteen nanometers, the width of a hundred or so average molecules. He called it a "lightsail," which had the double meaning of "light in weight" and "propelled by light." The only way to get a sail of such extreme thinness into space would be to manufacture it up there, but that was no problem. He'd come up with a plan by which you could do that — assembling, in orbit, everything from the foil itself to the finished spacecraft.

He outlined his scheme at a space conference he attended — this while an MIT junior. In the audience was one Mark S. Miller, a computer science major at Yale, and a man with a strong bent of his own for colonizing the planets.

"Eric had a concrete design for a machine that would extrude aluminum solar-sail panels out in space," said Miller. "Basically you'd vapor-deposit a thin sheet of aluminum onto some kind of waxy material which you'd then evaporate off. You'd be left with just a thin sheet of aluminum: it would be extremely thin, much thinner, much more delicate than anything you could hope to lift and unfold. By stitching together a large array of these extruded solar-sail panels, you'd make this twenty-mile-diameter sail that sunlight would bounce off of, and you could get to Mars in three months.

"I thought it was the most important paper at the conference," he added. "It did more to raise our notion of what was possible in space than any other single paper."

Later, for his master's degree, Drexler wrote up the specifications for his lightsail. In a thesis titled "Design of a High Performance Solar Sail System," he described his proposed "film sheet production system," which was in effect an automated, space-based sail-making machine. About twelve meters long ("well within the 18 meter length of the Shuttle payload bay"), it was an assembly line of rollers, coolers, heaters, spray nozzles, vacuum seals, motors, compressors, evaporators, and so on, all of it orchestrated to turn out a yard-wide strip of thin film at rates on the order of a linear yard per second.

The key fabrication process involved vapor-depositing a base layer of waxy stuff onto a moving belt, then spraying onto the waxy stuff an extremely thin film of aluminum. Finally you'd heat the waxy stuff to the point where it would vaporize off, leaving you at the end with nothing but the thin film. It was like spraying melted butter onto a cookie sheet, letting it harden, spraying some paint onto the hardened butter, and then melting away the butter: the only thing left would be the coat of paint.

Sections of the thin film would then be glued together (with rip-stop cuts inserted at regular spacings), until you had panels that could be hung from a framework, piece by piece, giving you a sail as big as you wanted — two miles across, for example, or even ten miles. The framework would be stabilized by a network of cables, and the whole thing could be rotated so that centrifugal force would flatten it out so it looked like a large silvery dish. Finally the payload would be attached to the sail by shroud lines, at which point the whole thing would be ready to travel. What you'd have at the end, Drexler said, would be a solar sail "with accelerations 20 to 80 times that of previously proposed, deployable sails."

His thesis adviser, Walter Mark Hollister, professor of aeronautics and astronautics at MIT, was an admirer of the scheme. "Eric was an unusual student," Hollister remembered. "He was a very gifted, far-out-thinking person. He should get credit for taking that vapor deposition idea and applying it the way he did."

At this point — it was May of 1979 — Drexler had two MIT degrees behind him, plus a short list of aerospace engineering proposals. He'd done theoretical work on producing refined aluminum from lunar ores, and he'd created a scheme for mining the asteroids. He'd done lab work with thin films, actually making some test models, once staying up all night to get them down to thicknesses of less than 100 nanometers. And he'd done some actual development work, with Gerry O'Neill, on a prototype "mass driver," an electromagnetic launch device for firing lunar ores into space. So at this precise stage of his career, anyone would have expected Eric Drexler to become one of the great geniuses of space travel, a sort of latter-day Wernher von Braun — which, in truth, was not much different from what he himself once might have thought.

And then, suddenly, everything changed.

"It was in the course of I would guess a few months that Eric moved from an interest in space to an interest in molecular engineering," Dave Anderson remembered. "We used to talk about space a lot and I remember that he was talking a little bit about molecular-engineering ideas during one visit to Cambridge, and then it seemed like the next visit he was already convinced that the road to space was through nanotechnology. He was already sold. Nanotechnology would lead to an economics of abundance, would lead particularly to the ability to design spacecraft that were light and easy to launch, and so forth. It would change things so much that space travel and space colonization would follow as a natural course out of the development of molecular engineering."

Shortly after they returned to classes in the fall of 1976, Chris and Eric began spending lots of time together. They went to parties and to the movies, and for a while they even worked side by side at a school snack bar, where they outdid each other making the thickest frappes.

"Everywhere else in the United States they're called milk shakes," Chris Peterson said, "but in Boston they're called frappes, and the goal of the students who worked there was to see how much ice cream you could cram into one of these things for your customers, who are all your friends and acquaintances. The management didn't like this, of course, because it was very expensive, but we always did our best to make those frappes as thick and big as you possibly could."

When they weren't working, studying, or attending classes, Chris would watch, a bit nervously, as Eric (this was in his more daredevil days) made bare-handed ascents up MIT's Alexander Calder sculpture. At least it was called a sculpture; mainly it functioned as a campus windbreak. Shortly after the formal dedication of the Green Building, which housed the Center for Earth Sciences, it was discovered that the ground-level doors couldn't be opened when there was anything more than the slightest breeze from off the river. In a masterstroke of advanced planning, the architects now put little glass baffles around the doorways, planted trees in aerodynamically strategic locations, and situated the massive Calder thing in the middle of the walkway that funneled the wind from off the Charles, after which the doors were openable in almost any weather conditions. But the sculpture soon became the school's technical climbing framework, and lots of students, including Eric, would climb to the top of it, execute a certain lay-back maneuver, and then reverse the procedure.

Mostly, though, Eric taught himself chemistry, as he was by now convinced that he had to take this molecular-engineering business to some kind of conclusion. If you were bent on constructing objects molecule by molecule, the one thing you had to have a pretty good working knowledge of was the science of how those molecules fitted together.

Chris Peterson, a chem major, helped him with this. She'd graduate with an award from the American Institute of Chemists ("In recognition of distinguished scholastic achievement, originality, and breadth of interest in chemistry and closely related fields"), and ended up tutoring him and loaning out her textbooks.

"He took all my chemistry books!" she recalled. "He borrowed them, anyway. We still have them today and they're all on his shelves. I had a complete library for him right there."

But now a new factor had emerged. Drexler had always had a knack for sizing up a situation, for following out a chain of reasoning to its logical consequences, to its bitter end, and when he did that with his molecular-engineering idea, he saw two distinct possible outcomes, one good and one not so good.

First there were all these vast and fabulous applications that the ability of working with molecules gave you: the medical miracles, the new materials, the suddenly cheap space stations, constructed automatically and at slight cost by this army of exquisitely programmed, self-replicating molecular machines.

All that, certainly, was on the positive side of the balance sheet. But those benefits would accrue only if everything worked out as planned, which was to say, if the molecular machines had been rightly programmed, if they executed their programs flawlessly, and if everything else went along like clockwork. But how likely was that to happen? How often did it occur in the greater world of science and engineering — and especially in the world of space systems and rocketry, which Drexler was most familiar with — that a long train of premeditated events went off without a hitch?

The answer was, not very often.

So then there was the other side of the balance sheet, the other possible consequences, which were not so rosy.

What if those molecular machines got out of control?

What if they got out of the box?

What if they started munching their way out of the lab, down the street, and . . . ?

"I was thinking about machines able to build copies of themselves that were highly nonbiological in their organization, not made of biomolecules," Drexler recalled much later. "Having done some reading in ecology, and having some understanding of the way the biological world works, it was pretty clear to me that it would be possible to build, not necessarily easily, but it would be possible to build a mechanism of the kind that could operate in the natural world, on abundantly available compounds, or perhaps a wide range of compounds, to build copies of itself. Something like that would be a lot worse than any plague or insect infestation you could think of, and in a limited case of awfulness such a thing could have a very broad ability to consume organic matter. Obviously there would be no predators, no ecological checks and balances. And so it could generally destroy the biosphere."

Destroy the biosphere. That was the other possible consequence.

Now Eric Drexler was nothing if not an extremely responsible individual, totally fastidious in his mental, moral, and physical habits, and suddenly there was this possible scenario in front of him whereby something he'd come up with as a benign and inviting engineering option — a realm of finely tuned, self-replicating molecular machines — suddenly had become monsters from outer space, a race of little wee robots running amok and trashing the planet in a world-class auto-da-fé.

The specter of which was not unprecedented: it had been the same vision, essentially, that had prompted the first wave of genetic engineers to hold back on their own creation, at least for a while. In 1973, during the early days of recombinant DNA, some of the experiments being proposed left a distinct bad taste in the mouth. There was the proposal for combining the DNA of a virus known to cause tumors in lab animals with the DNA of a virus that infected bacteria. Could the hybrid bacteria cause tumors in humans? If so, why take the risk of creating such a thing?

And there was the proposal for constructing a type of gene that could conceivably make bacteria resistant to certain antibiotics. Was that a wise move? However useful such a thing might be for research purposes, the thought of antibiotic-resistant bacteria did not sit well with some scientists.

So in 1974 a committee of the National Academy of Scientists called for a temporary ban on genetic-engineering experiments until people could come up with some safety guidelines. The next year there was a gathering of molecular biologists in Monterey, California — the so-called Asilomar Conference. Here, with an ocean full of natural life forms bobbing in the sea below, scientists debated what responsibility they had (if any) for the genetically engineered products they hoped to create. The next year, finally, a group at the National Institutes of Health drew up a list of working procedures. The rules were intentionally strict, but as further experiments were performed in safety, the initial guidelines were gradually relaxed.

Drexler was getting his molecular-technology ideas at about the time the NIH guidelines were coming out, and so he was quite aware of the possible dangers posed by an army of invisible artificial devices overrunning the planet.

"The package as a whole made me deeply ambivalent," Drexler said later. "It scared the wits out of me. There were these tremendous downsides. We were in the cold war at that point and there was a tendency for new technologies to be developed as weapons systems. I said, 'Well, this has terrifying possibilities, and I don't want to be responsible for advancing this process.' "

And so while he continued to think about the subject in private, and while he kept on discussing it with friends, he took no steps to publicize his idea.

"Eric was keeping his mouth shut so nobody would think of this any sooner than they had to," Kevin Nelson said later. "But if something is possible, somebody is going to do it, if there is a reason to do it. And with nanotechnology, there are plenty of reasons to do it."

By the fall of 1979, Chris had graduated from MIT and she and Eric were sharing an apartment. This was at 518 Putnam Avenue, in Cambridge, a two-bedroom walk-up on the third floor —

Chris's apartment, actually, which she took after getting a job at a local semiconductor company, Alpha Industries.

"I was involved in designing microwave diodes," she recalled. "At the time I joined the company, it was small enough that I could still be involved in many different areas, and I was also doing failure analysis, meeting with the sales staff, helping to figure out what should be in the catalog, and so forth."

Eric, meantime, had started working on a Ph.D. in aero/astro, which looked to be the logical next step, careerwise; at least it was the next step if he wasn't going to pursue the molecular-engineering idea on a formal level. On the other hand, a lifetime spent in aeronautical engineering really no longer appealed as it once had. He'd now seen molecular machines in the offing, and he'd had visions of the vast and fabulous things they could do. In light of which, the prospect of building space colonies by conventional methods — with human labor, with the old technology — suddenly, all of it seemed as if it belonged back in the Stone Age.

Then one night in the periodicals room — this was in November of 1979 — he picked up the latest issue of *Physics Today*, a special issue devoted to microscience. The lead article was an overview of the emerging field called "microtechnology."

The first sentence of the article was: "Just about twenty years ago, at the Christmas, 1959, meeting of the American Physical Society at Cal Tech, Richard P. Feynman gave a delightful talk, 'There's Plenty of Room at the Bottom.' "

Drexler had never heard of that talk (although he'd heard of Feynman), but the article went on to describe how Feynman had predicted that not so far in the future all kinds of stupendous feats of miniaturization would be possible. Supposedly, you'd be able to fit the entire *Encyclopaedia Britannica* on the head of a pin. You'd be able to build "miniaturized machinery," devices so small that they'd "manipulate at a nearly molecular scale." Feynman, the article claimed, had even talked about "placing individual atoms precisely."

Individual atoms?

This was a shocker to Eric Drexler, who'd been thinking all this time that the idea of working with atoms and building things up was his own private and personal creation.

But apparently it wasn't. Apparently he'd been anticipated, and to the tune of twenty years.

The *Physics Today* piece went on to say that a transcript of Feynman's talk had been published in a 1961 book, *Miniaturization,* edited by one H. Gilbert. Drexler looked it up in the card catalog, found it on the shelf, and started reading.

4

"Feynman Was Robbed"

It had been an after-dinner talk, the main event at the December 1959 meeting of the American Physical Society. This was an annual affair, held in the lull between Christmas and New Year's. The host institution this year was Caltech, but the banquet, because it was such an attraction — always a sellout — would be held off-campus, at the Huntington-Sheraton Hotel in downtown Pasadena. A crowd of three hundred, thoroughly wined and dined, is now looking forward to the evening's entertainment — or at least that's what they're hoping for.

"Being the organizer I'm always uptight," said Pacific Coast secretary Bill Nierenberg. "All I care about is that the speaker shows up and doesn't fall off the stage."

That wouldn't happen tonight. Nierenberg, who'd later become the director of the Scripps Institution of Oceanography, had invited Richard P. Feynman, Caltech's star physicist, to be the speaker.

"He accepted, which surprised me," Nierenberg said. "But we were exactly the same age, and we had in common the fact that both of us won the Putnam Prize in mathematics, which was a nationwide competition among math undergrads. You sat in a room and they gave you ten problems, five in the morning, five in the afternoon, and you broke your ass getting solutions to them."

In the banquet room, a giddy mood prevails. Feynman, although not yet the celebrity physicist he'd soon become, was already famous among his peers not only for having coinvented quantum electrodynamics, for which he'd later share the Nobel Prize, but also for his ribald wit, his clownishness, and his practical jokes. He was a regular good-time guy, and his announced topic for tonight was "There's Plenty of Room at the Bottom" — whatever that meant.

"He had the world of young physicists absolutely terrorized because nobody knew what that title meant," said physicist Donald Glaser. "Feynman didn't tell anybody and refused to discuss it, but the young physicists looked at the title 'There's Plenty of Room at the Bottom' and they thought it meant 'There are plenty of lousy jobs in physics.' "

Finally George Uhlenbeck, president of the American Physical Society, gets up and, amid sounds of silverware clattering, ice cubes clinking, and semi-inebriated physicists choking on their cigars, introduces the speaker.

A great burst of applause, and Feynman advances to the podium. After a moment he starts in with some apparently solemn and sobersides remarks about "opening up new branches of physics." Then he says: "I would like to describe a field in which little has been done, but in which an enormous amount can be done in principle. . . . What I want to talk about is the problem of manipulating and controlling things on a small scale."

Well, this made perfect sense. Miniaturization had been in the news for the last couple of years or so — ever since *Sputnik* had given scientists a new reason to think small. So many things now had to fit inside these satellites: guidance and control systems, power supplies, sensors and actuators, radios, computers, and God only knew what else. Whole industries were gearing up to pack all this stuff into the tiniest possible physical volumes.

But squeezing components into spacecraft was not what Richard Feynman had in mind. That was kid stuff. "That's *nothing*," he said. "That's the most primitive, halting step in the direction I intend to discuss."

What he intends to discuss is the question "Why cannot we

write the entire twenty-four volumes of the *Encyclopaedia Britannica* on the head of a pin?"

Ah yes! The entertainment portion!

"There was lots of laughter, as there would have to be," remembered Paul Schlichta, a Caltech grad student at the time. "The audience wasn't taking this seriously. They thought it was a big put-on joke."

As maybe it was.

"Let's see what would be involved," Feynman continued. "The head of a pin is a sixteenth of an inch across. If you magnify it by twenty-five thousand diameters, the area of the head of the pin is then equal to the area of all the pages of the *Encyclopaedia Britannica.* Therefore, all it is necessary to do is to reduce in size all the writing in the *Encyclopaedia* by twenty-five thousand times."

In visual terms, the total area covered by all the pages of the *Encyclopaedia Britannica* was a space roughly equivalent to half the size of a football field. Putting the encyclopedia on a pinhead, then, would require squeezing half a football field's worth of written text, some twenty thousand pages of densely printed prose and pictures, onto a tiny circular spot that was barely a sixteenth of an inch across.

"Is that possible?" Feynman asked.

It was *entirely* possible, he said, because at the required level of shrinkage — characters that were reduced by a factor of twenty-five thousand — the individual letters, numerals, and dots of the text would still be of finite and measurable size, some ten or so atoms wide.

"So, each dot can easily be adjusted in size as required by the photoengraving, and there is no question that there is enough room on the head of a pin to put all of the *Encyclopaedia Britannica.*"

At that scale, he went on, you could fit the text of the entire Caltech library — all of the words in all of the books in the whole collection — onto a single three-by-five-inch library card. At that scale, in fact, you could fit all of the world's important books — some twenty-four million volumes, he estimated — onto a few normal-size pages.

But that was not the best you could do — far from it. Indeed, it was next to nothing at all. It was just another "primitive, halting step" on the road to the bottom, the reason being that such a method of text storage utilized only two dimensions: it used only the surface of the pinhead. But there was lots more to a pinhead than that. A pinhead had a third dimension; there was a whole depth to it, a thickness, a veritable abyss of atoms stretching all the way from the top surface to the underside below, fully 1⁄32 of an inch away. That's where some real information could go, into that volume of metal.

If you reduced the alphabet to dots and dashes, and if you wrote out the text in dots and dashes using all the stratified molecular layers in the pinhead, all the way down, then you could pack an untold number of libraries into that space. In fact, you wouldn't even need the whole pinhead. If each separate bit of information was represented by a cube of metal that was five atoms on a side — 5 by 5 by 5, for a total of 125 atoms per bit — then you could fit *all* the information in *all* the world's twenty-four million books into a volume that was not only smaller than the head of a pin, but also smaller than a grain of sand.

"It turns out that all of the information that man has carefully accumulated in all the books in the world can be written in this form in a cube of material one two-hundredth of an inch wide — which is the barest piece of dust that can be made out by the human eye.

"So there is *plenty* of room at the bottom! Don't tell me about microfilm!"

"Feynman is like Fermi, you know," Bill Nierenberg said. "They're very different people with very different styles of delivery, but they have a way of presenting a subject that makes you believe everything they say. It's only after you leave the room that you realize you've been had."

Which at this point was what some in the audience were beginning to suspect. They were physicists, after all, and they'd earned all their hundreds of advanced degrees by reading books, studying books, going to the library for books, big damned fat square volumes with page after page of dense text, and here was a guy telling them that all of the world's important books could be stuffed into a grain of sand with plenty of room to spare.

That just didn't seem possible. But on the other hand, as Feynman now reminded his listeners, precisely such a feat — encoding countless bits of information in very small numbers of atoms — had already been performed in unaided nature, by DNA, the molecules of which embodied all the instructions necessary to manufacture a whole new human being.

"All this information is contained in a very tiny fraction of the cell in the form of long-chain DNA molecules in which approximately fifty atoms are used for one bit of information."

And so the next question was, if nature could do that, if it could reduce information to small, discrete numbers of atoms, then why couldn't we? Why couldn't we work with the atoms themselves — the fabled building blocks of creation?

"I am not afraid to consider the final question as to whether, ultimately — in the great future — we can arrange atoms the way we want; the very *atoms*, all the way down!"

As to how we were supposed to manage this miracle, Feynman was not at a loss. "Let me suggest one weird possibility," he said, going on to describe a series of machines each of which constructed an exact duplicate of itself, only smaller, until the last one in the series, absolutely the tiniest machine of all, could actually manipulate individual atoms. The machines would be controlled from the outside, by a human operator whose input movements would be duplicated by the machine on a reduced scale. Such an arrangement was like the remote-control mechanisms used in nuclear power plants, where the operator maneuvered a set of master hands that were connected to a set of slave hands which reproduced all of his original motions exactly.

You could perform some amazing tricks with such a system, merely by scaling down the ratios so that, for example, if the master hand moved an inch, the slave hand moved half an inch. Then, if the master hands went through motions of putting some physical object together, a doghouse, let's say, the slave hands would follow through with the same motions and assemble the identical doghouse, only a half-size version. And if the master hands constructed a full-size machine tool, such as a lathe, for example, then the slave hands would construct a half-size lathe.

Naturally, you could repeat the process on still smaller scales, making ever-tinier slave hands, until you reached the point at which the original master hands controlled an extremely tiny set of slave hands. So small would be these slave hands, Feynman said, that with them you could manufacture "one little baby lathe four thousand times smaller than usual."

Moreover, you could partition the process, distributing the operator's inputs to a whole bank of slave hands working in parallel. The big hands could control a hundred sets of tiny hands — or a billion — just as easily as one. You could make a billion tiny baby lathes, each identical to the next.

A billion tiny baby lathes!

What do you *do* with a billion tiny baby lathes? Why, you build a billion factories, of course.

"I want to build a billion tiny factories, models of each other, which are manufacturing simultaneously, drilling holes, stamping parts, and so on."

As to the question of what those tiny factories would make, Feynman's first answer was: computers. "Make them very small, make them of little wires, little elements — and by little, I mean *little.* For instance, the wires should be ten or one hundred atoms in diameter, and the circuits should be a few thousand angstroms across."

But in fact this atomic-scale mass-production line could manufacture more than computers. Indeed, it could manufacture virtually any device at all, and to atomic perfection: "If we go down far enough, all of our devices can be mass-produced so that they are absolutely perfect copies of one another. We cannot build two large machines so that the dimensions are exactly the same. But if your machine is only one hundred atoms high, you only have to get it correct to one-half of one percent to make sure the other machine is exactly the same size — namely, one hundred atoms high!"

That was Feynman's vision: atomic-scale computers and atomic-scale machines — the tools and devices of the Big World re-created in a matching universe below. By the time he'd gotten to the end of his talk, indeed, Feynman had not only opened up an entire new world to the reach of manufacturing, he'd opened it up to the prospect of massively distributed parallel-production techniques, and he'd even given some halfway reasonable hints as to how this dream could in fact be accomplished. He saw all of these events, furthermore, as predestined and inevitable, as developments "which I think cannot be avoided."

Still, he wanted to hurry them along any way he could, and so he offered some frank motivational inducements — "incentive funding," in latter-day terms.

"It is my intention to offer a prize to the first guy who can take the information on the page of a book and put it on an area 1/25,000 smaller in linear scale in such manner that it can be read by an electron microscope.

"And I want to offer another prize — if I can figure out how to phrase it so that I don't get into a mess of arguments about definitions — of another thousand dollars to the first guy who makes an operating electric motor — a rotating electric motor which can be controlled from the outside and, not counting the lead-in wires, is only 1/64 inch cube."

More of Feynman's jokes, apparently. A scale reduction of 1/25,000 was the precise scale at which the *Encyclopaedia Britannica* could be written on the head of a pin. That trick, even if it was possible in theory, did not seem to be just around the corner. As for the 1/64-inch motor, such a thing would be about the size of a period at the end of a sentence, and it was by no means clear that a motor of such dimensions was physically producible. So it seemed as if Feynman's money would be safe for a while.

But in fact he wound up paying off on both bets.

"From the moment the talk was over I had not the slightest doubt that the concepts were entirely feasible and that they should and would be pursued," said Richard Werthamer, who'd later become executive secretary of the American Physical Society. "No one rolled their eyes or wondered what the guy was smoking. The feeling maybe was like a great enlightenment, bordering on an epiphany. But there was also a strong element of imaginative surprise, like encountering a wonderful science fiction story where one keeps asking how the author came up with this amazing idea."

That was one reaction.

"I thought it was sort of interesting but I wasn't terribly impressed by it," said Robert Walker, one of Feynman's Caltech colleagues. "It was before its time and many of the examples were not ones where I could see an immediate use."

That was another reaction.

"People thought you could do *gedanken* experiments—thought experiments," said physicist Donald Glaser, who'd win the Nobel Prize for physics the very next year. "You could think about it, and you could compute what the limits might be, but I don't think anybody thought you could actually manipulate atoms. In those days the current folk wisdom was don't go too far or you'll run into kT—thermal vibration—and so you might as well not waste your time."

There it was already, the dreaded kT rearing its head!

"You had an audience of individual receptors in a sea of laughter," said Paul Schlichta, "and the two didn't notice each other. I was one of the profoundly impressed ones. I thought that something had been called to my attention that I should have realized long ago. Like all great ideas, it was 'obvious.' "

Well, it had been obvious to Eric Drexler, anyway.

When Drexler finished reading Feynman's talk he didn't know what to think. He was both elated and appalled. It was wonderful to find someone else thinking many of the very same thoughts that he'd had himself. It was wonderful to hear this described as a new field, one that had never been thought of before. And it was wonderful to see someone else, besides Drexler himself, taking seriously the whole idea of tiny computers, of atomically perfect, mass-produced devices.

But then there was the fact of having been anticipated to such a degree, and by almost twenty years. That was not wonderful. Feynman had spoken back in the Eisenhower days, back when the Physical Society banquet cost just $4.50 per person, back when

Drexler was just four years old. Suddenly, Eric's private new technology was old and public stuff.

Years later, Drexler recalled his emotions.

"Well, when you think you have something that is completely novel, and then to find some partial anticipation of it by a famous person, many years ago . . . you say, well, this could be read as being a strong anticipation, or it could be read as a weak anticipation, but it's kind of disappointing to have been anticipated at all.

"On the other hand, it's kind of useful to have a Richard Feynman to point to as someone who stated some of the core conclusions. You can say to skeptics, 'Hey, argue with *him!*'

"So it was a very mixed feeling. Mixed feelings on several planes."

But it wasn't only Feynman and Drexler who'd had thoughts of doing things with atoms. There was a way in which the idea was "obvious" to anyone who'd ever had it explained to them in kindergarten or the first grade how it was that atoms were the building blocks of matter. "Oh, what pretty atoms! Why can't we play with them?" And so forth.

But if atoms really were the building blocks, then why couldn't we actually build with them? Plenty of people had had that thought even within mainstream science, but few who did ever pursued it. And even among those who pursued it, few of them ever got very far.

There was the case of Arthur von Hippel, for example, who'd had such ideas in the 1930s, antedating even Feynman, who at that time was a teenager.

Born in Rostock, Germany, in 1898, von Hippel attended University of Göttingen during the 1920s when all the great guns in physics — Bohr, Pauli, Heisenberg, Dirac — were fetching up there to hammer out the new quantum mechanics. Later he arrived in Cambridge, Massachusetts, and became a professor of electrical engineering at MIT.

In the 1930s, von Hippel had gotten interested in the question of "the molecular events hiding behind macroscopic properties." If you could understand the properties of large-scale phenomena in terms of their molecular components, he reasoned, then you might be able to change those properties simply by altering the underlying molecular arrangements. This, he thought, was entirely "obvious."

"Nature designs everything from atoms," he said. "Hence, we should be able to design any kind of material with foresight if we thoroughly understood the periodic system in all its implications."

Later, in the 1960s, he introduced the term *molecular engineering*, and talked about "the building of materials and devices to order." The main advantage of doing this, of building things up from the atoms, was that you wouldn't have to accept brute matter in the form that nature gave it to you. Rather, you could tailor it to your own specifications.

"No longer shackled to presently available materials and characteristics, we are free to dream. Practically any structure which seems topologically reasonable can be made. Molecular designing allows us to realize Jules Verne's fantasies. The question is therefore not any longer what we can do, but what we want to do."

The only problem was, von Hippel had no real notion of how to go about any of this. He spoke in a language of metaphors and parables, of pleasing but vague images: "Designing with atoms is an occupation filled with curious anticipation, like gardening in a greenhouse: select the seeds, set the conditions of the surroundings, and things begin to happen."

Well, they'd begin to happen, all right, as soon as you put them in motion. But as for how to start the ball rolling, Arthur von Hippel was at a loss — as were all the others, apparently, who'd ever come up with the "obvious" idea of working with nature's building blocks. The "building blocks" notion was a metaphor, and it was best to leave it at that.

Feynman's talk was a nine days' wonder, portrayed by the media as the next new scientific sensation. *Popular Science* ran a condensed version under the headline "There's plenty of room at the bottom, says noted scientist as he reveals HOW TO BUILD AN AUTOMOBILE SMALLER THAN THIS DOT•"; other stories appeared in *Life*,

Science News, and *Saturday Review* ("Wonders that Await a Micro-Microscope"), which also reprinted an excerpt from the talk.

The complete text of "Room at the Bottom" was published in the February 1960 issue of the Caltech journal *Engineering and Science*. There was a picture of the author himself on the cover, all cheekiness and smiles. *Engineering and Science*, which was published by the alumni association, was mailed routinely to all Caltech graduates. One of them who received the issue was William H. McLellan, an engineer at Electro-Optical Systems, in Pasadena, maker of various types of silicon devices — lateral photocells and the like. McLellan, who'd gotten a B.S. in mechanical engineering from Caltech in 1950, read through the article with a mixture of fascination and disbelief.

"To me, as a practical man, I had some problems with it," he said later. "Like the idea of a very small machine programmed to make an even smaller one, and then a smaller one. You quickly run into dead ends because the pieces just mash into one another. Your tools, your techniques have to switch as you get smaller and smaller."

But McLellan, who had a good intuitive sense of what could and what couldn't be done with available technology, was attracted by Feynman's challenge, by the offer of $1,000 for an operating electric motor that was only one sixty-fourth of an inch on a side. McLellan looked at that and said to himself, "I could probably make one that size. But why? It'd be a lot of work." So he put the idea out of his head and thought nothing more about it until a few months later, when the June issue of *Engineering and Science* showed up in the mailbox.

"I read it, and then I realized that in all the subsequent issues nobody'd made one. I was sure it would be reported. So I said to myself, if I don't do something, this guy's offer is going to go unchallenged. So I set about to design the motor and make it."

He was pretty sure he'd succeed.

"Well, I knew some things," he explained later on. "For instance, I knew about wire that was one-half of a thousandth of an inch in diameter: that would be number fifty-six wire. I knew that you could buy it with insulation on it. I knew that you could bend it around a mandrel that was three-thousandths of an inch in diameter, without cracking the enamel. I knew exactly how to wind coils small enough so that four of them would easily fit in the required dimensions."

So McLellan took some of that number 56 wire, which was manufactured by the Sigmund Cohn Company, of New York or New Jersey or somewhere back east, for use in galvanometers, and wound it around each of four tiny iron pins set at the four vertices of a square base; they would constitute the stator, the stationary part of the motor. He then attached a tiny magnetized disk, the rotor, to an even tinier metal shaft — both of them together looked like a thumbtack — and set the shaft into a bearing centered among the four iron pins. In theory, when alternating current was sent through coiled wires of the stator, the rotor would turn in response to the induced magnetic field.

Because of the size of the parts, McLellan did all of the work while sighting through a $250 Bausch and Lomb microscope that he'd purchased specially for the occasion. After some two and a half months of this — much of it spent searching through the dust on his workbench for dropped, lost, or vanished parts — he had in front of him a functioning electric motor. It was one sixty-fourth of an inch square, had thirteen separate parts, weighed 250 micrograms, and had a drive shaft that rotated at speeds ranging from zero to about 2,000 rpm, depending on the frequency of the current sent through the lead-in wires.

He placed a call to Feynman, spoke with his secretary, and made an appointment.

By this time — it was November of 1960 — Richard Feynman had seen plenty of people walk into his office with boxes of various sizes from which they'd extract some rather hefty "small" motors. One fellow brought in a scale model V-8 engine that took up most of the palm of his hand. Somehow these people had missed the point just slightly, and so when Bill McLellan shuffled into Feynman's office with yet another large, brown box, Feynman mentally rolled his eyes.

What McLellan lifted out of the crate, however, was not a "small" motor. It was, instead, a microscope.

"Uh-oh," Feynman thought. "Nobody else brought a microscope."

The two men spent the next few hours putting the motor through its paces, switching it on and off, cycling it up and down through its speed range, and having great fun. It was a little weird, this machine, because no matter how fast the rotor spun, you couldn't hear anything.

"There isn't enough energy there to make any noise," McLellan explained. "We're talking about something a ten-thousandth of an inch in diameter, running at trivial surface speeds. There's not enough energy involved."

McLellan said good-bye to Feynman and went home. A few days later he got a $1,000 check in the mail.

In truth, it had taken Feynman a while to come up with the money. He'd made his two prize offers while he was still single, after all, and had never put any contingency funds aside.

"As a matter of fact," he explained later, "there was some consternation at home, because I got married shortly after that, and

William McLellan's 1/64-inch motor;
awarded the first Feynman prize, 1960. (*James McClanahan*)

had forgotten all about it. And when I was getting married I explained my financial position to my future wife, and she thought that it was bad, but not *so* bad. And it was about three or four days after we came back from our honeymoon that with a lot of clearing of my throat I had to explain to her that I had to pay a thousand dollars that I had forgotten, that I had promised if somebody made a small motor. So she didn't trust me too much for a while."

With the motor built and the prize awarded, everybody should have been happy, and that should have been that. But in fact it was a letdown for both parties.

"See, he was looking for new methods — that was the whole point of his talk," said Bill McLellan. "There's plenty of room at the bottom, but you've got to develop a new technology in order to exploit it. Maybe you could *grow* a coil. Imagine programming a plant that could grow a coil out of copper, and have insulation form on it. In other words, build it up atom by atom, nature's way. Do it efficiently: just build what you need, don't whittle away a whole bunch of stuff you don't need."

Feynman, too, felt a bit cheated. "I remember he was a little unhappy on the electric-motor thing," said Feynman's old grad student Paul Schlichta. "It was just a conventional miniature electric motor. He'd been hoping for an innovation, a new technology of some kind. I think he was thinking, in a way, of evaporated films, and so when the guy just simply made a conventional motor, that was a disappointment."

McLellan wound up feeling that he'd won the prize through a loophole, through Feynman's failure to specify in as many words that what he was putting up money for was not so much the motor itself, but the new technology that he'd thought (mistakenly, as it turned out) would be needed to build it.

"You could say that Feynman was robbed," Bill McLellan said. "He lost the thousand dollars and didn't gain a technology."

McLellan's motor was put on display at Caltech's physics department, outside Feynman's office, and across the hall from Murray Gell-Mann's.

"You could look at it through a microscope," Gell-Mann said of the motor, unimpressed.

And after that brief flurry of micromania, Richard Feynman's lecture, and Bill McLellan's motor, were ignored and forgotten for the next twenty years.

Barely had Drexler recovered from the shock of the Feynman piece when the other shoe dropped, making a big hole. This new indignity came in the form of a news item in the May 1980 issue of *Semiconductor International*, a trade journal. Chris had brought it home from where she worked because there was an article in it he might be interested in: "Protein Macromolecules Interface to Microcircuitry."

Eric, she knew, had been thinking of making computers out of protein molecules. He'd explained this to her time and again, he'd discussed it with his friends at the Space Habitat Study Group, but he'd never gone any further than that. And now, suddenly, others were having similar ideas.

Semiconductor devices, the article said, were now "approaching the dimensions of protein molecules." That was news, but that was not the important news. The important news was the claim that electronics firms, some of them, were now talking about actually making use of protein molecules, engineering them, fashioning them into component parts of computer circuitry.

This was too much! That had been more or less *his* thought, some three years earlier, and now others were out there retracing the same mental steps, coming up with the same ideas, ideas that he'd come up with on his own but then done nothing about because they were too dangerous, because they could lead to only God knew where.

Even worse was what one of these electronics developers, a James McAlear, had said about using genetic engineering as a fabrication tool. "Long range goals," he was quoted as saying, "include the use of genetic engineering to produce protein molecules for the specific application to microdevice fabrication."

"That's what I was thinking in 1976!" Drexler said to himself.

"I hadn't published anything in the interim," he recalled long afterward, "and suddenly there I was, three or fours years behind."

For a man who prided himself on his foresight, this was not one of life's happier moments.

5

Eternity and Clouds

When Kim Eric Drexler was born on April 25, 1955, at the naval hospital in Oakland, California, his father, an officer in the U.S. Navy Air Squadron, was off in Kodiak, Alaska, where his unit was stationed. He wouldn't see his son in the flesh for the next four months, but Allan Drexler knew exactly what his son looked like, because his wife, Hazel, took a Polaroid shot of the baby every single day of the week and sent it off to her husband, who papered the walls of his room in the Bachelor Officers' Quarters with them. He put up more than a hundred shots of baby Kim before he left, and he could see the child's progress quite clearly in the photos — especially between the first and the last.

"The first time I saw him I of course recognized him right away," Allan Drexler said later. "I knew what he'd look like, from all those pictures."

After he got out of the navy, the family moved to Lafayette, Indiana, home of Purdue University, from which Allan got a Ph.D. in speech pathology and clinical psychology. Then they were off to New Haven, where he took dual positions at Southern Connecticut State College, teaching, and at Yale, where he did clinical work. Plus, he taught a for-credit college course on public television. The show was a hard sell with his son, however, because Allan was up against the cartoons.

"On Saturday mornings we'd watch television, Eric and I. I'd sit with him and watch the cartoons — he loved 'Bullwinkle' — and we'd flip over to me every once in a while."

They moved again, to Cincinnati, where Allan went on to much success and minor fame in his field, coauthoring a textbook, heading up the Cincinnati Speech and Hearing Center, and developing a Ph.D. program in speech pathology and audiology for the University of Cincinnati. On weekends, he'd take his son sailing on the boat he kept on the Ohio River.

He also took Eric to the 1964 New York World's Fair, which the boy went nuts over.

"He was in awe of it," his father recalled years later. "He couldn't get enough of it. He wanted to stay even longer and I think we stayed two, maybe three days."

The World's Fair was something, all right, especially to a nine-year-old boy. Both Allan and Hazel used to read science fiction

Kim Eric Drexler, age 9, at 1964 New York World's Fair. (*Allan Drexler*)

stories to him — they'd taught him how to read, in fact — and now stretched out in front of him was the most futuristic, otherworldly sight he'd ever seen: buildings that were soaring curves and flying disks; saucer-shaped things that looked like they'd just arrived from Mars; there were cubes, rings, cones, wedges, spheres — but all of them were actually buildings, "pavilions" — many of them hanging in thin air, it seemed like, or supported by slender cement buttresses that could not possibly hold them up, but did. And a monorail curved past on the outskirts, just like in the science fiction strips.

The exhibits themselves were just as phenomenal. All the big companies — General Electric, IBM, Ford, Coca-Cola — had outdone one another with simulations of the near, far, and indefinite future. There were computers all over. There were robots that walked, talked, and looked like real people. There was a simulated moon-walk. And at the General Motors "Futurama" pavilion there were these moving chairs that sped you through the next hundred years or so and then, when they popped you out at the end, there was an attendant who handed you a lapel button that said, "I have seen the Future."

Well. Who could resist the spell of it all — the magic, the aura, the high-tech allure? Certainly not Kim Eric Drexler.

Back home in the real world, unfortunately, Allan and Hazel were getting divorced — not that this was all that bad a tragedy. Eric remained with his mother, who was always the more science-minded of the two parents. "She's the one who gave him that very good mathematical, scientific background," Allan said.

Hazel Drexler, née Gassmann, was most definitely Eric's connection to the outer cosmos. She'd gotten her own cosmic outlook at the age of four, as she and her family were driving to the West Coast. This was before the age of motels, and so when the family stopped for the night they all slept out under the stars. One night, she recalled, "We saw shooting stars. Oh, it was wonderful, just wonderful! Magical! This was out in the desert of Arizona or New Mexico. Very clear sky, no city lights to interfere, dry desert air — it was just great!"

Later, in high school, Hazel took every available science

course. She was junior-class president, graduated at the top of her class, and was valedictorian. Then, at Penn State, she majored in mathematics, graduating with the single highest grades in the Liberal Arts College — plus honor roll, Phi Beta Kappa, and everything else. From there she went on for a master's degree in the subject, and then taught mathematics at Penn State and, later, at several other colleges.

But she wasn't just a head-in-the-clouds theoretical type: she was mechanically inclined and shrank from no project. "I've wired a house," she said. "My first husband and I built a house and I did all the electrical wiring, including wiring the box where the main line comes in."

She also fixed her own cars. One time when the fuel pump broke, "I went out to find a used fuel pump to put on this thing, and I couldn't find one that was right but I found two fuel pumps, and the top fitted one place, the bottom fitted the other, so I put 'em together and it worked."

During the Second World War Hazel test-ran aircraft engines, wrote technical reports for the Wright Aeronautical Corporation, in New Jersey, and later worked at the U.S. Naval Torpedo Station, in Newport, Rhode Island, as an engineering inspector and draftsman. She was, in all, a highly independent woman — "too independent to be an ideal wife," she said. (By 1992, indeed, she'd been married three times.)

Anyway, after her first divorce she and Eric moved to Denver, where her brother, Andy, would serve as a father figure to the boy. Like Hazel herself, Andrew Gassmann was a Penn State mathematician with a craze for astronomy. He taught high-school science in Castle Rock, Colorado, helped start the Gates Planetarium in Denver, and taught at the University of Denver planetarium for twenty years.

"Andy would come up for supper," Hazel remembered. "He'd take Eric to astronomy meetings, take him to 'star parties,' take him up in the mountains to see the stars."

All of which left quite an impression on the young Eric: the mountains, the stars — and also the clouds. Off to the west were the Rockies, and in the afternoons, especially in the summer, there were these incredible cloud buildups — big, vertically developed, white cumulus clouds that seemed to loft straight up for miles into the clear, cold blue. They were brilliant, mesmerizing, exhilarating presences.

After a year or so in Denver, Hazel, who'd gotten a Ph.D. of her own in speech pathology and audiology back when the family lived in Cincinnati, took a job at the Oregon College of Education (later Western Oregon State College), in Monmouth.

This was a town of about five thousand in the Willamette Valley, a place awash in trees, fields, and crops, and with snow-covered mountain ranges off on both horizons: the Coastal range low toward the west, the Cascades a little higher toward the east. Here, Eric did the rest of his growing up.

"Eric's education, at least until he went to MIT, was at least fifty percent on his own," his mother recalled. "He wasn't interested in grades, he wasn't interested in clubs. He mostly learned on his own, in his own room, reading books, doing projects."

There was the lunar agriculture project, for example.

"I was interested in space development, and if you're going to live in space you're going to need to know how to grow food," Eric said years later. "There had been consideration of farming on the moon, and some people had pointed out that on the moon we have two weeks of sunlight and then two weeks of darkness. One proposal for keeping the plants going during the two weeks of darkness was to have nuclear power plants generate electricity, and have the lights on all the time. And I looked at that and I said, 'If you cool them down, why can't you just keep the plants in darkness for two weeks?' "

So he did the experiment. He took two identical flowerpots, planted alfalfa seeds in each, and grew them for a while in ordinary sunlight — the usual twenty-four-hour day/night cycle. Then he put one pot in the closet, the other in the refrigerator, and kept them there for two weeks. When he took them out, the refrigerated plants were still living.

"The ones that had been in the warm, dark closet had gotten long and spindly, and they died when I put them in the sun; the ones that had been in the refrigerator did just fine. So I concluded that lunar agriculture was perhaps more feasible than people had thought."

He also kept a trained spider — or one that appeared to be trained, anyway. Eric would constantly move the branch it lived on so that the web would be spun in a certain pattern. He had a microscope and did drawings from it — extremely detailed pen-and-ink sketches of biological cells.

"At that point, when I saw those drawings, that's when I realized that this person is not going to need me very much," his father recalled. "He's gonna make it on his own."

By the late high-school years, Eric was a thin wisp of a fellow with a mustache and goatee, and with a slightly distracted air, as if his mind was off somewhere else. Like his mother, he was very much the nonconformist. When other kids were wearing these absolutely identical peace symbols — the leather-thong variety, dangling from the neck — Eric created his own personal version. With short pieces of masking tape he made an outline of the peace symbol on his chest and then went out and sat in the sun.

"In two or three days he took the tape off and he had a great tan . . . with a peace symbol on his chest," said his father. "A lot of people would buy a peace symbol, or draw one. He got a peace-symbol tan."

He also stopped using the name Kim — which, after all, was this default-value, unisex name his parents had saddled him with.

"We came across 'Kim' in a list of names," his mother recalled. "Allan and I both liked it, and since we didn't know whether it was going to be a boy or a girl, we said we'll call him Kim regardless."

But he kept the initial, and so he'd now be "K. Eric Drexler," as if he were a bank president or head of the FBI. Anyway, his friends no longer teased him about his name.

"Eric was always something of a social outcast, basically because there were very few people who understood him," said Dave Anderson, one of his best friends from the Monmouth days. "He was regarded as your classic nerd taken to the extreme. I remember people teasing him about talking in lingo because he used words that were much bigger than what most of the people understood — because he talked about science and space and so forth."

Eric and Dave would take these tremendous all-day bike rides — not so much for exercise or scenery, but for the sake of visiting a decent library. The nearest one was twenty miles to the south, in Corvallis, on the campus of Oregon State University. There was an actual engineering library there, not just the baby collections they had locally in Monmouth.

"We tended to talk a lot about the cosmos and the stars," Dave Anderson remembered. "I had a telescope, a four-and-a-quarter-inch Newtonian reflector, and we'd occasionally set it up on my roof or out in a field and look at the planets. I can remember particularly one time we stayed up, I believe, all night, seeing almost all of the planets. We started with Mercury and then Venus, and we worked our way around. We were almost ready to put the telescope away when we saw a bright star coming up in the east, and we decided to point the telescope at it. It was a lot of work pointing this telescope at anything — it had a very poor mount — but we finally managed to point it over there, and for both of us this was our first view of Saturn.

"To this day I consider Saturn to be the queen of the heavens," he added. "It's just the most gorgeous view you can have."

In 1972, during his senior year at Central High, by which time he'd begun to take physics and calculus courses over at the college ("I got early release for good conduct," he said), Drexler happened upon a book that would transform his worldview and shake him to the core. The book was *The Limits to Growth: A Report for the Club of Rome's Project on the Predicament of Mankind.*

Written by four MIT professors, *The Limits to Growth* purported to be a scientific forecast of the world's future. On the basis of a computer program that plotted the rise and fall of five interrelated quantities over time — population, agricultural production, natural resources, industrial production, and pollution — the authors claimed to have discovered that unless wholesale economic and social changes were made forthwith, human civilization would collapse in very short order.

"If the present growth trends in world population, industrialization, pollution, food production, and resource depletion continue unchanged," they said, "the limits to growth on this planet will be reached sometime within the next hundred years. The most probable result will be a rather sudden and uncontrollable decline in both population and industrial capacity."

Everything hinged on their "World Model," a computer simulation of nothing less than . . . the entire world. The simulation was depicted in the book by a two-page flow diagram showing the five interrelated quantities, with the relationships among them modified, qualified, and hedged about by a sizable helping of fudge factors. There were "social adjustment delay" factors, "lifetime perception delay" factors, "auxiliary variables," et cetera and so on, all of them connected up in a complicated arrangement of crisscrossing lines, feedback loops, manifold interlinkages, causal dependencies, transitional states, nodes, "valves," and more. The thing was a masterpiece of complexity and commotion, on the basis of which the authors set forth a bunch of italicized apocalyptic warnings:

"The basic behavior mode of the world system is exponential growth of population and capital, followed by collapse. . . .

"We can thus say with some confidence that, under the assumption of no major change in the present system, population and industrial growth will certainly stop within the next century, at the latest."

There were some two hundred pages of this.

Matters looked dire indeed after all those feedback loops, delay factors, and italicized proclamations of global doom. Still, there was a solution. All the world had to do, the authors said, was to *stop growth.* That was their formula, their medicine, their universal panacea. Population and wealth must just stop their incessant and ugly expansion, after which all would be sweetness and light forevermore.

This was not offered as a suggestion, recommendation, or a point for the reader's humble consideration. No. It was offered, instead, as an ultimatum, as a command, as a requirement. The authors were the planetary police stamping out the evil growth forces. Issuing their decrees in the charming linguistic idiom "We require that," they proclaimed: "We require that the number of babies born each year be equal to the expected number of deaths in the population that year."

They required, further, "a state of global equilibrium." They required *"deliberate checks on growth."* They required a stoppage of excess capital formation, after which "additional production is not allowed."

The world economy must be dipped in formaldehyde, preserved in amber. That way, it would live.

Well, none of this sat too well with K. Eric Drexler. Since early youth he'd looked forward to a future of human betterment, progress, and growth. He looked forward to increased diversity, opportunity, advancement — to the further flowering of the human species. And now, suddenly, that whole shining vision was dimmed, wiped out, replaced by a Stalingrad of coercive restrictions and controls, planetary checks and restraints. All at once God's green earth, which ought to have been a free and open place, had been turned into a vast global prison.

The worst of it was that he himself couldn't see a single point at which the authors had gone wrong. As much as he hated their conclusions and prescriptions, he could see no flaw in their overall analysis. Taking their reasoning as a whole, it did indeed appear that continued growth meant ultimate collapse, that short-term progress meant long-term mass misery.

In a note to himself which he made at the time, Drexler wrote: "*Limits to Growth,* although I know that it is not a prophesy or a timetable has affected my point of view deeply, because I have yet to read a convincing argument against the substance of its conclusions. It has placed limits of time and material extent on my mental model of Earth's future, has helped me set personal priorities by helping to determine what is and is not important, and has depressed me for several months."

The irony was that Drexler and the *Limits to Growth* authors were, in a way, kindred souls. Both he and they operated from within the perspective of an extremely long-term, global, macrocosmic viewpoint.

The *Limits* authors had made a special point about this, having gone so far as to create, in their ostentatiously scientific fashion, a "human perspectives graph," in which they plotted out the time spans of people's interests against the number of people having those interests. Most people, it was clear from the graph, had narrow interests that stretched across relatively short time scales. "Only a very few people have a global perspective that extends far into the future," the authors said.

But it was precisely such a perspective that Eric Drexler himself possessed. His outlook, in fact, was more than merely global; it was universal, it was ageless, it was eternal.

There'd been the time, for example, when his mother's cat died, which happened while Eric was still in junior high.

"He died in my arms," his mother recalled. "And I cried. I felt terrible. And when Eric got home — he was still Kim then — I told him the cat had died and that I was crying. He didn't cry, though, even though he was fond of the cat. He said, 'Well, I don't cry about death.' But he was trying to explain to me what made him cry, and he said, 'One day I was looking out the window and I saw those cumulus clouds. And I thought how thousands of years in the past there were cumulus clouds, and that millions of years in the future there will be cumulus clouds. And then I cried.'

"It's this eternal perspective he has," she added. "It had something to do with eternity, and time. That touched him very deeply."

But now these *Limits to Growth* authors, people who manifestly shared the same global perspective, they'd peeked into the future with their semi-mystical computer simulations and high-tech tea leaves and had seen nothing but Bleak House ahead.

This was unacceptable. And in fact, as Eric soon came to realize, it wasn't even necessary. The whole *Limits* scenario had been predicated on a single central assumption: that the world beneath our feet was the only world we had. This was stupid. It was like a group of hunter-gatherers, realizing they'd pretty much exhausted the one tiny patch of ground they were standing on, deciding they had to reduce their population instantly or else starve to death. Whereas all they needed to do was to move on.

The human race, Drexler thought, was in much the same situation: *There was an entire universe out there!* Even on earth our mines had barely scratched the surface of the planet, so in all reason how could the *Limits* folks talk as if the only resources available were those we had access to *today?* They'd taken this "global" perspective of theirs just a bit too literally; there was no reason to think solely in terms of "the globe." We'd already been to the moon and back and had even returned with some rocks. The asteroids were out there — a whole solar system, in fact! How could the *Limits* crew not see this point?

By the middle of his senior year in high school, Drexler knew what had to be done. The door to space had to be reopened, and he'd help reopen it. He'd show people what was really possible if you went out to the asteroid belt and to the planets. He'd even show them how to do it, as he'd already developed a few ideas on the subject.

But first there was an education to get, which now loomed as an all-important desideratum. There was a human future to be saved. A world to be reclaimed.

Earlier, his father had suggested that Eric take a year off from school before he went to college. Go to Europe, his father had told him; bang around, experience things. "See the world for a year," he'd said. "What's the hurry?"

And Eric had even considered it for a while. *Travel!* (But secretly, he found travel boring.) *See the world!* (Who cares?) But then *The Limits to Growth* came along. He could do without this year of "European studies" or whatever.

Nevertheless, just to be on the safe side, to placate the gods of travel, Eric took a short mental journey. He wrote up an elaborate account of it, a travelogue of sorts, for his senior-year project. He called it "My Trip to Nepal."

"He made it all up!" his father said. "Absolutely made it up. I was hysterical when I read it because he had details in it like you wouldn't believe. He was talking about going into the capital. He was talking about the mountain range and what it was like, the view, what the people were like.

"My guess is, it was a gestalt way of saying, 'I'm gonna close that loop. I'm gonna write this paper as if I did it, so I won't have to worry about it anymore.' It was sort of closing a loop in his life, that trip."

Drexler applied to Stanford, Cornell, and MIT, but the last was where he really wanted to go. The other two colleges, fine as they were, had a definite academic bent to them, they leaned toward pure science, toward understanding for its own sake. MIT, by contrast, bent over in the other direction, toward applied science and engineering. You went to MIT not so much to know the way the world was, but to do something about it, to make things, to build.

On his application to MIT he wrote: "Over the past year I have developed a strong interest in the possibility of establishing human civilization in space. Given the energy, materials, and conditions known to be available or probably available in the asteroid belt, it appears that it is a favorable location for the existence and growth of an industrial human society. Since the asteroid belt contains more available energy and material than the Earth by at least an order of magnitude, the limits to the growth of such a society would not be reached until it contained many billions of people at a high standard of living. The creation of such a society, if possible, would be the moral inverse of genocide, and would put mankind beyond accidental destruction or final collapse.

"Over the next four years I would like to make a study of at least one pattern for an asteroidal civilization, outlining energy and raw materials sources, extraction processes, and production of the necessities of life, all based on near certainties and conservative estimates. Then, if the results look as promising as the thinking I have done so far, I would like to study what needs to be done, that is, find the minimum of equipment, personnel, and money necessary for the founding of a self-sufficient colony capable of growth, and find ways of making such a project politically feasible."

A year after writing those words, Drexler, now a nineteen-year-old MIT freshman, was up at the lectern of Gerry O'Neill's First Princeton Conference on Space Colonization, presenting his paper about mining the asteroids.

One thing Richard Feynman's talk did for Eric Drexler was to make him think yet again about publishing his own ideas on the subject. Feynman had said that working with atoms was "a development which I think cannot be avoided." Eric had agreed with that 100 percent: sooner or later people were going to figure out how to do things with the atomic building blocks; it was only a matter of time.

And then, after the *Semiconductor International* piece about making some industrial protein molecules, well . . . suddenly it seemed as if the time had come.

The problem, in Eric's eyes, was that neither Feynman nor the *Semiconductor International* piece had mentioned any of the potential hazards in developing such powerful new capabilities. Everybody seemed to be looking only at the bright side of the picture. In Feynman's whole talk, not once did he refer to the risks or dangers of building tiny invisible machines; it was as if there weren't any such risks. Nowhere did the *Semiconductor International* story say anything about possible accidents or abuses. To the contrary, it spoke in glowing terms about the prospect of implanting protein-based integrated circuits into human brains, thereby giving people "an increase in effective intelligence of a large magnitude." It talked about "the addition of senses that humans do not directly possess." And so on.

Not that Drexler thought that such things were impossible, undesirable, or even unlikely; as far as he was concerned they were probably doable. The trouble was, those people saw only one side of the story. They were in gee-whiz mode. They were starry-eyed.

Where was there any responsible thinking about all the various and sundry things that could go wrong? Where was there a realization that Truly Bad Consequences were a distinct possibility?

"And at this point," Drexler recalled, "I got some inkling of the fact that having a long-term perspective was a little rare. But this was an area where it was very important for the ideas to be articulated in a way that included the long-term perspective, so that we weren't just heading toward a cliff. And I concluded that I had an obligation to publish these ideas, and therefore be in a position to articulate the consequences.

"The way to have influence on technology development is to play some active leadership role," he added. "Because then you're involved, you're in the middle of things, and people will tend to listen to what you have to say about it."

That, at least, was the hope.

There were some problems, of course, when it came to publishing what amounted to a brief for a whole new technology, especially one that started at the molecular level and built up, creating molecular computers, cell-repair machines, and all the other miracle devices. The main problem was, how did you make such a technology believable? It was one thing to walk into a roomful of your space-cadet friends, members of the Space Habitat Study Group, and start spouting chapter and verse about putting molecular machinery to work building you some truly glorious (but cheap) interstellar spacecraft. It was quite another thing to deliver the same sort of message, or even a drastically toned-down version, to a bunch of solemn and grim physicists, engineers, and aggressively straitlaced hard-science types. You couldn't simply say to such people, "Look, there's this great new technology off in the wings, it can perform all manner of incredible feats, wiping out disease, poverty, hunger, and aging — and by the way, you should start preparing for it now because otherwise it might just possibly, ahem, get out of hand and, you know, destroy the biosphere."

No, you couldn't start off by saying just that. You had to approach the precipice gradually, little by little, preparing the way with an elaborate consideration of the various technical hurdles involved, which, to the chorus of cynics that was undoubtedly out there, would be daunting enough all by themselves. Every scientific and technical advance was at first greeted by a vast battalion of naysayers, pessimists, and professional doubters.

But the hurdles confronting molecular technology seemed to be quite real. There was the chicken-and-egg aspect of the whole thing, for example. How was it possible to create tiny machines before you had machines that were already small enough to do the work involved? You couldn't build such machines until you had tiny tools, but you couldn't fabricate tiny tools unless you had some tiny machines to make them with. This was a Catch-22 from which there appeared to be no escape.

And that was only the beginning. There was the problem of controlling those machines once you'd created them. There was the question of supplying power to them — how was that little feat to be accomplished? There was the issue of radiation damage and errors caused by thermal effects. And there was the problem of getting a macroscopically meaningful product out of such tiny, invisible, submicroscopic entities. Such apparent obstacles (which is all they ever were in Drexler's mind) could be overcome, of course, but there was no getting around the fact that you had to solve them one by one and lay out the answers for all to see.

As it was, Feynman had already come up with a way to escape the chicken-and-egg situation. He'd build his tiny machines through a process of successive approximation, using bigger machines to build smaller ones, seriatim. That was the equivalent of starting with a macroscopic egg and whittling away at it until you'd reached the level of molecules. It was a top-down approach.

Drexler's approach was precisely the opposite. Instead of starting at the top and whittling your way down, you'd start at the bottom and build up. You'd begin with nature's own molecular devices: DNA, proteins, lipids, and so on, reprogramming them to do your bidding. You wouldn't have to create the machines, you'd use the machines that were already there — in particular, the mechanisms of protein synthesis — to build new, better, and more sophisticated molecular devices.

Proteins, after all, were nature's own wonder molecules. Not themselves living things, they were nevertheless what living things, by and large, were composed of. Much of the human body's structural material — its muscles, connective tissues, the cell walls, and so on — all these things were made up of proteins. Much of the body's functional elements were likewise proteins: hemoglobin was a protein; hormones were proteins; enzymes were proteins. The proteins, in short, were the body's all-purpose building material and molecular operating equipment.

There was nothing like them on the macrolevel. Protein molecules were long chains of amino acids, hundreds or thousands of them linked together in a single line. There were exactly twenty different naturally occurring amino acids, and each type of protein was composed of a different and unique sequence of them.

The naturally occurring proteins were highly functional devices. Some of them, the enzymes, were shaped in such a way as permitted them to physically wrap around a given, smaller molecule, holding on to it so that a chemical reaction could take place, such as the addition or removal of an atom. Other proteins were complex enough to act as molecular motors, such as those that turned the flagella of a bacterium.

The most remarkable feature of a protein molecule, however, was the fact that the sequence of its component amino acids caused the molecule to "fold" into a given shape. That was the term biologists used to describe the way in which an amino acid string behaved once it was placed in water and let go: the string kinked up, curled around, twisted, crimped, and "folded" back upon itself in a highly individual and specific fashion. The shape of the fold was determined by the precise order in which the different constituent amino acids were distributed along the chain.

Each separate amino acid had its own distinct molecular personality. Some were hydrophilic, meaning that they had a chemical affinity for water; others were hydrophobic, tending to avoid water as much as possible. Some were large and bulky, others were small; some rigid, others flexible. When placed in water, a protein's hydrophobic groups would curl inward so as to shelter themselves from contact with water, whereas the hydrophilic groups would billow outward so as to maximize water contact. These and other such mechanisms caused a distinct ordering of amino acids to assume a specific derivative three-dimensional shape: one particular sequence led to one particular structure, reliably, every time.

It was exactly this tendency of a protein to fold up in a predictable and invariable fashion that Drexler wanted to exploit. A protein was already a molecular machine. Why not simply modify one or more proteins so that they'd fold up into the shapes and structures you wanted them to have? You'd decide what final shape you wanted the folded-up protein to take, and then you'd figure out what amino acid sequence was required to produce it.

It sounded easy: all you needed to do was to come up with the right amino acid sequences. Once you'd figured out what those sequences were, then standard genetic-engineering techniques would enable you to get biological cells to manufacture them. Placed in water, the resulting amino acid chain would fold up into your desired shape, and *voilà!* There was your designer protein molecule, waiting and ready for use.

There was only one small hitch in this whole optimistic scenario. The hitch was that no one knew which amino acid sequences led to which folded-up shapes. In biological organisms, the relationship between the two had been established over the eons by the process of evolution by natural selection. If a given sequence of amino acids wasn't useful for a given purpose, it wasn't selected for and didn't get repeated. The human body manufactured some sixty thousand different types of proteins, and did so automatically. But although scientists knew what the underlying mechanisms were, they didn't yet know how to take a given sequence of amino acids and predict the way they'd fold.

The task of designing new proteins, then, was akin to sending a coded message without knowing what the code was; or decrypting such a message without knowing the key; or translating from one language to another without knowing either language, and without having access to the proper dictionary. A precise one-to-one relationship existed between the elements of the first domain and the associated elements of the second; you just didn't know what that relationship was.

Nevertheless, that in essence was Drexler's plan for creating a race of molecular machines: sequence the right amino acids together and thereby create a marvelous new protein to order. Create enough of those proteins of the right size and shape, and they'd assemble themselves into a workable device — into the molecular machine of your choice.

Neither he nor anyone else knew exactly what orders to give, but that was simply a matter of engineering design and practice — trial and error, if need be. Sooner or later science would figure out which distinct sequences led to which distinct shapes, and at that point molecular machinery would be yours for the asking.

6

Richard Comes to Chris and Eric's

It took him a solid year to write the article that, in published form, would run to all of four pages. Not that working on it was a full-time activity; halfway into the writing he switched temporarily to what was, for him, an entirely new field: Xanadu hypertext. He'd heard about it from Mark Miller, at one of the Princeton space conferences.

"We talked about his lightsail proposal and all that," Mark Miller recalled, "and then he asked me what I was working on. So I told him about Xanadu. I was expecting the reaction that we normally get when we tell people about Xanadu, which is they don't really get it at first, or they get a small piece of it. It just takes a while for it to sink in, they have a lot of questions and things.

"But from Drexler what I got was, his eyes got real wide, he put his hand up to his face and he said, 'Do you know how important that is?' It was the quickest appreciation of how important the thing is that I've ever gotten in over ten years of explaining Xanadu to people."

Xanadu hypertext was the result of taking an engineering approach to human knowledge. The knowledge-base, when you thought about it, was an incredibly scattered and fragmented thing. It existed in no one location; rather, it was here, there, and everywhere: in people's heads, in libraries, data banks, archives, museums, laboratories, and a hundred other places. Some of it was cataloged, some of it wasn't. From an engineering point of view, such a method of storing knowledge left a lot to be desired. Getting to the truth about a given subject was no mean feat. A standard literature search, for example, involved not only coming up with a list of relevant books and articles but physically locating them on library shelves, in "the stacks."

They were a nightmare, those stacks. They were organized according to principles that generally baffled the human user, especially when he or she discovered that half the items under one subject heading were off in special collections, or over in the architecture library, or at the school of engineering, which for your convenience was located across campus or in another part of the city. Here a stack, there a stack, here a little microfiche — a routine lit search was sometimes enough to drive you crazy.

The reason was that while knowledge was cataloged according to rigid plans — the Library of Congress classification, or the Dewey decimal system, or whatever — neither reality itself nor human interests mirrored such schemes in the slightest degree. Reality was complete and whole and came all together in one big chunk, whereas human interests were arbitrary and diverse and followed no generally predictable lines or patterns. When the diversity of human interests ran up against the monolithic organizational scheme of the average library, the result was a momentary stoppage of the intellect.

The first person who both realized the problem and tried to do something about it was Theodor ("Ted") Nelson, who'd come up with the original hypertext notion in 1965. Ted, "writer, showman, generalist" (according to himself), was working on a graduate degree in philosophy when he was struck by the essential disharmony between writing, which was sequential, and reality, which wasn't.

"Your body is not sequentially interconnected," he said, nor were your ideas. "The structures of ideas are not sequential. They tie together every which way."

Well, then, why not come up with a method of writing and reading which reflected both the true nonsequential structure of things as well as the ideas that were meant to embrace them?

Nelson was working on a broad philosophical approach to this — a philosophy he called General Schematics — when, at about the same time, he took his first course in computer programming.

Which changed everything. Computers, he realized, were models of flexibility: they could be programmed to do just about any conceivable task. Texts stored in computer memory could be indexed and cross-referenced in such a way that finding your way through a bunch of related documents was easy. The computer could take the same batch of data and view it in six million different ways. It could trace out all kinds of unexpected themes and connections; in a matter of seconds it could find linkages that it would take human beings countless hours or weeks to discover.

"Computer storage and screen display mean that we no longer *have* to have things in sequence," he said. "Totally arbitrary structures are possible."

Soon Ted Nelson was imagining ways in which computers could allow you to view texts in parallel, simultaneously, side by side on the screen. This "parallel-textface" function would be highly useful when you were following the course of argument on a controversial issue — a situation in which there was lots of back-and-forth complaint and commentary. With parallel textface you could read one document and while in the middle of it switch to some other document on the same screen, so that you could compare the two, checking a fact, verifying a quotation, seeing what the other guy said. You could see if the author had ever replied to a critic, and, if so, you could see whether that critic had made any response to the author. Because it went beyond the essentially passive functions of mere text, Nelson called this dream-system of his "hypertext."

"Hypertext has a lot of the virtues of both conversation and of writing, while leaving aside a lot of the flaws of each," Mark Miller explained. "In writing you have the virtues that what you write is persistent, that you can really think through things before you express them, and that you can express things in some length, so you can express complex ideas. The problem with writing is that if you express something complex and I express a complex answer to it, people reading your document are generally unexposed to my response to it. With conversation, on the other hand, if there are several people in the room and you make some claim and I make a counterclaim, then everybody who is in the room who heard your claim can also hear my counterclaim.

"The important thing is not so much the counterclaim itself," he added, "but whether you have a response to my counterclaim. In conversation, people can hear the absence of a response to my counterclaim, whereas in a writing medium it's so hard to find a counterclaim, a criticism, that to be able to visibly see the lack of a response to a criticism is almost completely impossible. But it's something that's trivial with a group of people talking in a room."

Anyway, when Ted Nelson tried to implement these functions on an actual computer he ran up against a blank wall. For some reason, hypertext was just exceptionally hard to program. But he wrote up an account of it in his 1974 self-published book *Computer Lib / Dream Machines*, in which he whimsically compared his system to the mythical land of Xanadu, a far-off dreamy place that you can never quite reach. He spoke of "Xanadu hypertext," and "The Xanadu Project," which, eventually, would implement it.

A year later, in 1975, Mark Miller went to Yale, fell in love with computers, read *Computer Lib / Dream Machines,* and suddenly he, too, was hooked on hypertext.

"It was about five pages in *Computer Lib,*" Miller recalled. "I read this stuff and thought it was real exciting, so what I decided to do for my class project for this introductory programming course I was taking was to implement hypertext."

Eighteen years later, he was still trying to implement hypertext. By this time, though, he'd learned the awful truth about computers.

"Ted has this wonderful saying about computers that's ideal for understanding the weird state I got into," Miller recalled. "He says that the first thing to understand about computers is that with them, 'Everything is possible, nothing is easy.' When I first got into computers, I was overwhelmed with the first part of that, and I hadn't realized the second part yet."

Nelson, as it happened, lived in Swarthmore, Pennsylvania, near Miller's own home town of Philadelphia, so one summer Miller went over to see him.

"At that time I kind of apprenticed myself to Ted. In almost all ways, I learned vastly more about computers through my apprenticeship to Ted, which was three summers altogether, than through four years of being a computer science major in college."

Soon, Ted Nelson, Mark Miller, Roger Gregory, and some other sharp-witted programmers who'd been enthralled by the hypertext vision had rented a house in King of Prussia, outside Philadelphia, where they hoped to make the dream come true. The major problem, one that stymied everyone else who'd worked on the project, was getting the computer to keep track of all the various thematic links that were possible among an ever-burgeoning number of documents.

"The issue is the combination of fine-grained links and history and version tracking," said Miller. "We thought we had a solution to all these problems in 1980. We had a set of algorithms and data structures that were clearly significantly beyond the corresponding state of the art in computer science."

And then finally, just when they thought they had the problem licked once and for all, Eric Drexler became involved.

After he'd talked with Miller at the space conference, Drexler had become increasingly attracted to Xanadu hypertext. For one thing, the molecular computers he'd been thinking about since 1976 would be the absolutely perfect medium for a hypertext system. You could fit an entire library in your pocket; you could call up whatever information you needed as and when you needed it.

But he was interested for another reason, too, which was to maximize the chances of his molecular-engineering ideas being accepted by the wider scientific community. The world of science, he well knew, was as tradition-bound and blinkered as any other when it came to what was truly new and unorthodox. If you had a viewpoint that ran flat contrary to conventional wisdom, you were a flake, a crackpot, a crank. Or you were peddling "science fiction."

But with a hypertext system keeping track of the debate, people would have to be a bit more guarded in their claims. They'd have to be a bit more responsible, because as soon as their statements were entered into the record, then those attacked could respond, also on the record, allowing anyone following the dispute to judge where the truth lay. Molecular engineering, Drexler thought, could stand up well enough to criticism, just so long as people could see both sides of the issue.

"During the summer of 1980," Mark Miller recalled, "we brought Drexler within our proprietary boundary after having him sign a nondisclosure agreement. We explained what our internal proprietary data-structures were, and we taught him the general theory."

But the general theory, Drexler saw quickly, was quite seriously defective.

"Drexler understood our stuff very deeply and was able to spot some really bad flaws in it," Miller said. "The group of us had sort of been telling each other why the theory seemed plausible, why it would work, and we had talked ourselves into thinking that what we had come up with would work a lot better than it actually would. Drexler spent a lot of effort explaining to us where the actual flaws were, and convincing us of that."

Not until three years later, when Drexler came up with his own data structure, was there a further breakthrough. Drexler's structure was unique in the annals of hypertext because, unlike everything else that had been tried, it worked.

The whole experience left Mark Miller with a new regard for Eric Drexler.

"Drexler's work on hypertext was in an area where I'm a personal expert," he said. "It's an area that since 1988 I've had full-time involvement in a commercial project that Autodesk is putting millions of dollars into, building the system up. And it's just completely clear that Drexler was able to spot fatal flaws in systems that other people who were experts in the field thought were plausible designs; and then he came up with an alternative solution that he'd actually thought through, and that actually does work.

"He's an exceptionally careful engineer," Miller added. "Drexler is the most conservative engineer I've ever met. He's careful to distinguish between those things which he simply believes to be true, and those things that he knows he can make a solid case for. And he's very careful to publicly promulgate only the latter."

By any standard, "Molecular Engineering: An Approach to the Development of General Capabilities for Molecular Manipulation" was an unusual document. It wasn't every day that a scientific paper proposed a new technology, but that was exactly the claim of this one. "In this paper," Drexler wrote, "I will outline a path to this goal, a general molecular engineering technology." Published in the September 1981 issue of *Proceedings of the National Academy of Sciences,* the piece was low-key and restrained to the point of dullness — until, that is, you got the full drift and import of its message.

"Development of the ability to design protein molecules will open a path to the fabrication of devices to complex atomic specifications," he wrote. "This path will involve construction of molecular machinery able to position reactive groups to atomic precision. It could lead to great advances in computational devices and in the ability to manipulate biological materials. The existence of this path has implications for the present."

The garden-variety scientific paper was a statement of results achieved: it told you what had happened in the lab, or in the observatory, or in the particle accelerator, or whatever. This was almost a rule of scientific publication, the purpose of which, after all, was to communicate some new discovery about the world. Except when proposing new observations or experiments, a scientific paper was not a forecast of what *might* happen or what *could* happen or what *would* happen if everything else worked out right.

But this is precisely what Drexler's *PNAS* paper did: it told you what could, would, and should occur at some indefinite point in the future. All of it was couched in the subjunctive, or more precisely in the subjunctive disguised as the timeless present: Here is what *can* happen.

"Gene synthesis and recombinant DNA technology can direct

the ribosomal machinery of bacteria to produce novel proteins, which can serve as components of larger molecular structures."

"Intermolecular attraction between complementary surfaces can assemble complex structures from solution."

"Molecular assemblages of atoms can act as solid objects, occupying space and holding a definite shape. Thus, they can act as structural members and moving parts."

"Sigma bonds that have low steric hindrance can serve as rotary bearings able to support $\approx 10^{-9}$N. A line of sigma bonds can serve as a hinge."

There were claims as to what *ought* to be possible or what *should* happen: "The engineer should be able to design proteins that not only fold predictably to a stable structure (sometimes) but that serve a planned function as well."

What all of it meant was that this was an *engineering* paper, not a scientific paper per se. It would be a long time before people understood this point, but even when they did, many of them regarded its claims as fantasy.

Which, at the time, was not unreasonable. Practically the whole of Drexler's argument rested on the possibility of protein engineering — an undertaking that, to put it mildly, was not yet a going concern. Some experts, indeed, doubted that it ever would be. Protein folding was just too complex and unpredictable a phenomenon for you to be able to design proteins and then expect the ribosomes to manufacture them to order.

But supposing that it *would* be possible — itself a great leap of the technical imagination — Drexler's argument required a second such leap immediately.

The second leap was the supposition that a bunch of different proteins could be designed to fold up in mutually congruent ways, so that when placed in solution they'd assemble themselves into a working molecular machine. Just let them loose, and they'd magically — or at least spontaneously, automatically, without any further guidance or direction — fit together like jigsaw pieces, forming, at the end, a tiny operational device.

Strange as that might have sounded, it was not in fact so great a leap as it appeared. Self-assembly, after all, was a known and standard natural process. Crystals were self-assembling; viruses were self-assembling; the ribosomes themselves were self-assembling molecular structures. Clearly, if nature could avail herself of the process, then so could human beings. It was a matter of designing the parts correctly.

There were some further jumps to be made after that, but none so spectacular as the first two. The next was that with this primitive, first generation of self-assembling molecular machinery you could "build a second generation of molecular machinery."

These second-generation devices could be quite more adept than the first. They could approximate tiny industrial robots, and could be equipped with "molecular arms wielding molecular tools." Those molecular tools could then manipulate individual atoms and molecules, placing them precisely, thereby assembling any and all further machines, structures, or materials — fabricating anything else, basically, that was physically possible. Or, as Drexler phrased it in his calm, restrained, conservative-engineer's fashion: "The class of structures that can be synthesized by such methods is clearly very large, and one may speculate that it includes most structures that might be of technological interest."

Any argument rested on certain presuppositions, and in this case two were noteworthy. One was that atoms could be treated as real, mechanical, Newtonian objects, more or less as if they were tiny marbles. This flew in the face of the reigning orthodoxy, which held that atoms were indefinite fuzzballs or vague nothings bobbing about on a sea of quantum uncertainty and thermal noise. But if that were true, Drexler reasoned, then stable molecules would not be possible, and neither would atomically precise molecular structures such as DNA or proteins. Indeed, if atoms were so very uncertain in their positions, and if thermal vibration was all that bad, then biology itself would not have been possible.

Which led to the second presupposition, that molecular biology was a proof-of-concept of molecular technology. To Drexler, if the one was possible, so was the other. Molecular engineering was only an attempt to duplicate, through design and intention, what biology had already accomplished by natural means. To illustrate

Table 1
Comparison of Macroscopic and Microscopic Components

Technology	*Function*	*Molecular Example(s)*
Struts, beams, casings	Transmit force, hold positions	Microtubules, cellulose, mineral structures
Cables	Transmit tension	Collagen
Fasteners, glue	Connect parts	Intermolecular forces
Solenoids, actuators	Move things	Conformation-changing proteins, actin/myosin
Motors	Turn shafts	Flagellar motor
Drive shafts	Transmit torque	Bacterial flagella
Bearings	Support moving parts	Sigma bonds
Containers	Hold fluids	Vesicles
Pipes	Carry fluids	Various tubular structures
Pumps	Move fluids	Flagella, membrane proteins
Conveyor belts	Move components	RNA moved by fixed ribosome (partial analog)
Clamps	Hold workpieces	Enzymatic binding sites
Tools	Modify workpieces	Metallic complexes, functional groups
Production lines	Construct devices	Enzyme systems, ribosomes
Numerical control systems	Store and read programs	Genetic system

the depth and extent of the parallels between the two realms — the mechanical and the biological — Eric included in his article a table of comparisons.

In the face of all these analogies between mechanical and biological devices that performed the same function, how could anyone deny that man-made molecular machinery was possible?

"To deny the feasibility of molecular machinery," Drexler wrote, "one must apparently maintain either (*i*) that design of proteins will remain infeasible indefinitely, or (*ii*) that complex machines cannot be made of proteins, or (*iii*) that protein machines cannot build second-generation machines."

But none of these claims was in fact very plausible. Protein design would happen sooner or later. Complex machines (animals) were already made of proteins. And as for more primitive protein machines being able to build more complicated, "second-generation" machines, well, developments in synthetic organic chemistry provided evidence that that could happen. Indeed, so carefully had Drexler rehearsed his overall case for molecular machinery that he seemed to have a response to every objection, an answer to every question.

If the question was "How were these machines to be powered?" the answer was: "Conformation-changing proteins (such as myosin) can serve as sources of motive power for linear motion; the reversible motor of the bacterial flagellum can serve as a source of motive power for rotary motion."

If the question was "Can macroscale objects (such as people) control these tiny nanoscale machines?" the answer was: "As present microtechnology can lay down conductors on a molecular scale (10 nm) and molecular devices can respond to electric potentials (through conformation changes, etc.), such devices can be controlled by human operators or macroscopic machines."

And if the question was "Won't these machines suffer damage from thermal vibration and cosmic rays, making them error-prone and unreliable?" he had an answer to that, too. One of the paper's reviewers — Philip Morrison, in fact — had asked him precisely that question.

"It was a valid point for him to raise," Drexler recalled later. "I hadn't really done a thorough job of analyzing or arguing the reliability of the assembly operations before that. It made me go back to the biology literature, and when I did I found an instance of an organism that did DNA replication with an error rate of one in 10^{11}."

Which was good news indeed for his complex molecular systems. "As engineers commonly design systems to function reliably with many more failed components than 1 in 10^{11}," he wrote, "such an error rate seems no barrier to the construction of quite complex devices."

The one thing Drexler said the least about in his *PNAS* paper, ironically, was the very thing that had kept him from publishing for all these years: the possibility of molecular machines getting out of hand and "destroying the biosphere."

"Part of my thinking had been that you don't necessarily want to emphasize the adverse consequences because a lot of them involve military weapons," Drexler explained. "On the one hand there's ignoring things and not knowing that they're coming, and on the other hand there's talking about it and focusing on it so much that it becomes the center of activity, and it becomes a self-fulfilling prophecy. And I sort of oscillated back and forth between the two."

And if truth be told, he probably also realized the sheer folly, from a public-relations standpoint, of describing a futuristic new technology, one with which you could build "most structures that might be of technological interest," and then adding, almost as an afterthought, the warning that, "Oh yes, this stuff could wipe out the world in short order if you're not careful." At any rate, the only thing the *PNAS* piece said about the fateful subject of "dangers" was: "Those concerned with the long-range future of humanity must concern themselves with the opportunities and dangers arising from this technology."

On the whole, Drexler's first formal publication on nanotechnology was decidedly on the unusual side. It reported plans, not results; it was a work of engineering rather than a work in science proper; it proposed a way of viewing atoms that many would regard as naive, as a throwback to the bygone prequantum days; it outlined a new type of chemistry, one in which bonds were established by the forcible, mechanical placement of molecules, rather than by random diffusion; and it spoke of various new molecular devices, including computers, completely automated "molecular scale production systems," and a cadre of biological cell-repair machines that would work wonders with the human body. The overall program it described, furthermore, was one that itself depended on still another new technology, protein engineering, which unfortunately did not as yet exist.

It was hard to know what to make of the piece. It was the kind of argument, after all, that couldn't really be refuted. Who could say that any of it was "wrong"? How could you deny what was, at bottom, a prediction, a forecast, a bunch of claims that various things could, would, and "can" happen? How could you evaluate what amounted to a bunch of hopes?

Still, it was rather amazing that a twenty-six-year-old grad student at the MIT Space Systems Laboratory was advancing, in however conjectural and speculative a fashion, specifications for a whole new technology, one that would manipulate atoms and molecules, and manufacture, at labor costs that "can approach zero," anything that was physically possible.

Drexler's hope was that the readers of his paper would be so excited by the possibilities it described, and so convinced of the fundamental rightness of his argument, that they'd rush out and begin work.

Which was not quite the way it turned out.

Eric's paper was submitted to the National Academy of Sciences on June 4, 1981. Two weeks later, he and Chris got married.

The ceremony took place at a Unitarian church in Williamsville, New York, just outside Buffalo, Chris's home town.

Chris's father was a mechanical engineer at Union Carbide; her parents had divorced when she was about eight or nine, after which she, like Eric, had been raised by her mother. Chris had always been good at math, and in high school took part in math contests, often winning prizes and plaques.

"I thought I was hot stuff," she recalled later. "That's why I went to MIT. That's why everyone goes to MIT, but when they get there they find out, well, they're just one of many others who think they're hot stuff."

Eric Drexler, though, really was hot stuff, at least in Chris Peterson's view. "He was really smart," she said. "He was very smart, very serious, he clearly had important goals, he was working very hard. He was a lot more focused and determined in what he was doing and a lot smarter than most of the other people there. I liked that. I have always been attracted to people who are very, very, *very* smart."

Chris took lots of science courses at MIT and ended up majoring in chemistry. "That was my default major. I liked math, I liked chemistry, I never did like physics. My physics teachers weren't very good. Even at MIT I had crummy physics teachers, and I guess that makes a big difference. One January I went to Caltech to visit a friend there, and we sat in on a very informal question-and-answer session led by Richard Feynman."

This was Feynman's course "Physics X," which he taught for nearly twenty years.

"He did it once a week, like Tuesday nights or something," Chris recalled. "You'd go and he'd answer physics questions for the undergraduates. Anyway I sat there and I listened to him explain physics and I thought, 'If I could just learn physics from this man!' He made everything so clear. He was just an amazing guy."

Neither Chris nor Eric were religious in the least, but for the sake of family harmony they had a conventional church wedding, which they survived. Eric looked atypically rakish in a dark blue suit, tie, long hair, and river-gambler mustache; Chris, a smiling angelic presence in an off-white satin gown, her short hair ringed with flowers.

About sixty friends and relatives showed up for the reception, including Eric's father, who came from Annapolis, and his father's father, who was then in his seventies. Eric's mother, Hazel, came all the way from Livingston, Montana, her home at the time. Most of the Xanadu crowd was also on hand.

"The Xanadus came because they wanted to brainstorm with Eric," Hazel recalled much later. "They and Eric were up all night long, and then they'd sleep during the day. Mark Miller would use the bed that I had slept in, which was Chris's mother's bed, and then I'd use it again at night. We just kind of switched off."

Conversation at the reception tended toward the technical, even for the jokes. "It was during Drexler's wedding reception that we coined the idea of the Douglas Hofstadter joke," Mark Miller recalled.

Douglas Hofstadter, author of *Gödel, Escher, Bach,* had just recently started writing a column for *Scientific American.* He was obsessed with the notion of self-reference, the property of a sentence's referring to itself, and many of his columns had discussed the subject in excruciating detail. One time he even ran a self-referential short story (written by David Moser, another self-reference fanatic), which began:

> This Is the Title of This Story,
> Which Is Also Found Several Times in the Story Itself
>
> This is the first sentence of this story. This is the second sentence. This is the title of this story, which is also found several times in the story itself. This sentence is questioning the intrinsic value of the first two sentences. This sentence is to inform you, in case you hadn't already realized it, that this is a self-referential story, that is, a story containing sentences that refer to their own structure and function. . . . This sentence is introducing you to the protagonist of the story, a young boy named Billy. This sentence is telling you that Billy is blond and blue-eyed and American and twelve years old and strangling his mother.

And so on like this for the next three and a half pages.

"We kept trying to invent ever-better Hofstadter jokes," Miller continued, "and then Eric came up with the one that I think is the best of the genre, which is: 'Why did Douglas Hofstadter cross the road?' Answer: 'To make this joke possible.' "

Anyway, the newlyweds left for a week's honeymoon in the White Mountains of New Hampshire, staying at a place that was near both Thorn Hill and Thorn Mountain. One morning they decided to hike up Thorn Hill.

"It took an awfully long time," Chris remembered. "But the view was great. We found out later it was Thorn Mountain."

They returned home to the Putnam Avenue apartment in Cambridge. Chris was now a product manager at Alpha Industries, where she'd be for a total of five years. Eric, meanwhile, was enjoying some minor renown at MIT. Earlier that year, on four consecutive Tuesday nights in January, he'd given a new IAP course, "Coming Technical Revolutions." As described by the course listings, the lectures were a digest of Drexler's current research interests:

Jan. 6
"Lightsails"
We can get around the solar system for about one-thousandth the cost of using today's chemical rockets. Goodbye, limits to growth.

Jan. 13
"Xanadu Hypertext"
Writing knowledge on pieces of paper fragments it horribly. We need hypertext, 'The Wings of Mind.'

Jan. 20
"Molecular Engineering"
A path lies open to a technology able to build devices to complex atomic specifications. With it, we can remake the world or destroy it.

Jan. 27
"Curing Frostbite"
With molecular engineering, we can develop a cure for extreme, long-term frostbite. For today's society, the implications are stunning.

Curing Frostbite?

This was Eric's restrained, conservative-engineer's way of referring to the topic of suspended animation, the process of freezing the dear departed in order to prevent tissue decomposition, in hopes that at some future time the party in question might be brought back to life. Prior to coming up with molecular machinery and the cell-repair devices it made possible, Eric had considered the notion to be . . . well, science fiction. When people froze to death out in the wilderness, after all, they didn't magically spring back to life after thawing.

But molecular-scale cell-repair devices quite changed the picture. With an army of them coursing through your body, frozen tissue repair ought to be easy. He'd said as much even in the *PNAS* piece, where, without speaking of anything so weird as "reviving the dead," he described a few rather surprising things that such devices could, should, and "can" do.

"Molecular devices can characterize a frozen cell in essentially arbitrary detail by removal and characterization of successive layers of material (atomically thin layers, if desired). With frozen tissue, knowledge of normal structures and analysis of frozen structures should permit quite accurate reconstruction of the nature of the tissue before freezing."

And so in his new MIT lecture series he laid out chapter and verse of these and other "stunning implications" of his little molecular marvels. There was a pretty substantial turnout for these lectures, including, of all people, Carl Feynman, Richard Feynman's son.

"At the beginning of the molecular-engineering talk," Drexler recalled, "I outlined the basic idea of using positional control of atoms to build molecular structures, and I asked if there was anyone in the audience — at this point I was trying to find out whether these ideas were new (it seemed a little bit surprising to me that they were, because they seemed to me to be so obvious) — and so I asked whether anyone in the audience had ever heard of any discussion of an idea like this. And a hand went up in the back, and the guy said, 'Well, a talk by Richard Feynman in 1959.' And I said, 'Yes, that's the first citation in my paper.' Anyone else? No. So I went on with the lecture, but when I talked with the guy afterward it turned out to be Carl Feynman."

Marvin Minsky and his daughter Margaret were in the audience, too, and soon word-of-mouth about this remarkable "Eric Drexler" phenomenon was spreading beyond the narrow borders of MIT. When Margaret attended a conference in Washington, DC, for example, she met Paul Trachtman, who was an editor at *Smithsonian.* Naturally she told him about Eric Drexler — "He has ideas, so many ideas!" and so on — and soon Trachtman had invited Drexler to write them up for the magazine. Which is how it happened that *Smithsonian* ran two feature-length pieces by Drexler, one on lightsails ("Sailing on sunlight may give space travel a second wind") and one on nanotechnology ("Mightier machines from tiny atoms may someday grow").

The latter piece was illustrated by an artist's conception of a "machine built, atom by atom, from protein parts." This showed your basic modern nuclear family — man, woman, child, and dog — gaping in awe at a mechanical inner works of some sort, complete with a conveyor belt carrying a line of disattached human and gorilla heads.

Here as elsewhere, nanotechnology would defeat many an artist.

Every so often, Chris and Eric would throw a party. These were Chris's idea, initially, but Eric grew to love them.

"He liked intellectual stimulation," said Kevin Nelson, who attended the gatherings. "He wanted to bounce ideas off people, he wanted to hear their ideas, he had fun with ideas. They were important."

Eric would draw up written invitations for these "Future Parties" — they'd say something on the order of "Come to our party and talk about the future. Bring a friend" — and there'd be a map with directions at the bottom. One party, in April 1981, celebrated the launching of the space shuttle.

"They were on Saturday night," Dave Lindbergh recalled, "and the place was incredibly packed full with people. Very warm. These were all MIT-type people, all kinds of weird people, very interesting conversations going on."

There were the usual health foods: popcorn, potato chips, corn chips, and drinks.

"There was this aluminium cookpot with a skull and crossbones on the side labeled ETHYL," Kevin Nelson recalled. "It was punch, homemade punch. They didn't make it in the lab, they just went down to the package store."

One night, Richard Feynman came to Chris and Eric's party.

Eric had become friendly with Carl Feynman after that first lecture, had invited him to the parties, and one time, lo and behold, in through the door walks Carl Feynman, *et père.*

"I introduced him as 'Richard,' " Drexler recalled. "I

was a little too embarrassed to say, 'This is Richard *Feynman.*' "

Eric and Richard, though, did not exactly hit it off.

"I proceeded to encourage him to hold forth on various matters of physics, and so on," Drexler recalled. "We talked about the *PNAS* paper some, and generally he indicated that, yeah, this was a sensible thing. I don't remember a whole lot of terribly specific content, but at one point I was talking about the need for institutions to handle some of the problems it raised, and he remarked something along the lines of, institutions were made up of people and therefore of fools. That's a paraphrase but it captures the gist of it. My response was that I regarded myself in such matters as a fool, thank you. We're all human beings, me included, but we try anyway."

Kevin Nelson had a much better time talking with this 'Richard' character.

"I didn't know who he was," he recalled later. "Some kid's father. I was feeling claustrophobic so I got up and went over to Eric's bookshelves to look at his books, and there was this other guy doing the same thing. And I thought, 'Ah, a grown-up!' "

The others had been talking about using Eric's cell-repair machines to halt human aging, making possible life spans on the order of thousands of years. Nelson asked Feynman what he thought of the idea.

"He said he had a hard time accepting the notion that people would *want* to live for five or ten thousand years," Nelson recalled. "But then I told him how rare it was for people to turn down extraordinary lifesaving measures even when they were quite sick. At that time this was very much in the news, how people in the hospital were using techniques to stay alive for a few more weeks or months, even at enormous cost. But if this cell-repair stuff worked, I said, then you wouldn't be keeping somebody sick alive, you would be keeping somebody in good health alive, which changes the picture. If suicide was this rare and if turning down these lifesaving measures is this rare now, I found it highly unlikely people would forgo them in the future.

"Anyway, we had a long discussion. I finally went back and sat down, and somebody next to me said, 'So what did Feynman want?' And I said, 'Who?' He said, 'Richard Feynman — you know, Nobel Prize . . . ?' And I thought: 'Geez!' "

At which point Nelson thought back to what Feynman had said about Eric's molecular machines.

"That's simple stuff," Feynman had said. "It's obvious. Why doesn't he work on something difficult?"

7

Brother Eric's Nanotech Revival

In the technical literature, Eric's *PNAS* paper would be cited in the form "(Drexler 1981)." There were no such citations for a while, but then in 1983, two years after it was published, it was mentioned in both of the world's two top science journals, *Science* and *Nature.* Both articles were about protein design, and in one case the author had latched on to the fact that when it came to the question of protein folding, Drexler had come up with a key insight, one that was probably original. This was that there was not one but in fact two protein-folding problems: one was that of *predicting* the shape of a fold; the other was that of *causing a fold to happen.*

Drexler had noticed, apparently for the first time, that the two problems were logically separate and distinct, that you could solve one of them without at the same time solving the other. The first was a scientific question, the second a matter of engineering. For purposes of developing a molecular technology, the engineering problem was the only one you had to solve: you had to be able to *cause* proteins to fold up in certain ways.

As for how to do that, Drexler had advanced two suggestions in the *Proceedings of the National Academy of Sciences.* One was to stack the cards as best you could. Although accurate fold prediction was not yet possible, a few rules of thumb did exist. It was known, for example, that hydrophobic amino acid groups tended to curl themselves inward, to avoid water, whereas hydrophilic groups did the reverse. Eric suggested that by reference to this and other such rules of thumb you could try to induce, provoke, or inspire a given fold, even if you couldn't in fact "predict" it in the strict scientific sense of knowing in advance that it would happen. "Through the use of strategically placed charged groups, polar groups, disulfide bonds, hydrogen bonds, and hydrophobic groups, the engineer should be able to design proteins" — design them, that is, in the sense of causing them to occur as intended.

His second idea was a little less fancy. Basically it was to experiment, to try things out and see what worked. If Thomas Edison, as the story went, could perform thousands of experiments before coming up with an operational light-bulb filament, the budding protein engineer could do likewise with amino acid sequences, trying out different likely combinations until finding those that produced the desired result.

"Engineers (in contrast to scientists)," Drexler had written, "need not seek to understand all proteins but only enough to produce useful systems in a reasonable number of attempts. . . . Even a low success rate will lead to an accumulation of successful designs."

Drexler's basic idea, then, was to strong-arm various folds into existence, creating designer proteins through a combination of engineer's "feel" for the amino acid building materials and a general willingness to experiment with them. However unpromising such a proposal might have sounded, by 1983 other scientists were actually following it up.

In "Designing Proteins and Peptides," which appeared in the January 20, 1983, issue of *Nature*, author Carl Pabo noted that in the *PNAS* piece "Drexler speculates that it should be possible to design novel proteins and that such proteins could provide 'a general capability for molecular manipulation.' He points out that it may not be necessary to solve the protein-folding problem before we are able to design proteins."

Pabo spoke of Drexler's having suggested an " 'inverted' approach," in which "rather than starting with an amino acid sequence and then predicting the conformation of the folded polypeptide, one starts with a conformation of the backbone and then picks an amino acid sequence that should stabilize it."

The "backbone" of a protein was an internal skeleton of carbon and nitrogen atoms on which the various distinct amino acid groups were hung like pendants on a necklace. Drexler had not used the term *backbone* (nor the term *inverted approach,* for that matter), but the concept was there nonetheless in his idea of using "strategically placed charged groups," and the like, to induce a given pattern of folding.

In Pabo's view, the "inverted approach" of starting with "a pre-folded backbone" made sense and might work. "In fact," he said, "the inverted approach may simplify proteïn design even after the folding problem is solved."

Three weeks after Pabo's piece came out, a new article, "Protein Engineering," by Kevin Ulmer, was published in *Science.* Protein engineering was not only possible, said Ulmer, but on the verge of happening. Scientists already knew how to manipulate DNA, and did so commercially: that was the whole sum and substance of the biotechnology revolution. Soon it would be possible to program genes so as to produce any number of made-to-order proteins. Citing Drexler's 1981 *PNAS* piece, Ulmer endorsed its main conclusions, claiming that successful protein design would be "the first major step toward a more general capability for molecular engineering which would allow us to structure matter atom by atom."

Subsequent discussions of protein folding often took Drexler's "inverted" approach as their starting point. In 1987, for example, the Yale biologists Jay Ponder and Frederic M. Richards wrote in the *Journal of Molecular Biology*: "Drexler (1981) has suggested that it may be useful to express the problem as an inverted question: 'What sequences are compatible with a given structure?' . . . The present work proposes an elaboration of Drexler's (1981) suggestion."

And so on. All of it was flattering enough. Indeed, the idea of two Yale molecular biologists elaborating on an MIT aero/astro grad student's suggestion for making designer proteins that would assemble themselves into primitive molecular manipulators — well, the circumstance was slightly mind-boggling to contemplate.

Still, no engineered proteins had yet appeared on the scene, much less any molecular manipulators made up of them.

The logical next step for Drexler to take was to produce a full-blown account of his molecular-engineering scheme, a technical document that fleshed out the whole story in chapter and verse, with all the clinical details. That was the obvious thing to do, anyway, if he wanted to convince the greater science and engineering world that molecular engineering was a real prospect and not just his own private fantasy.

Not that this ought to take all that much convincing. As time went on, the less of a fantasy molecular technology seemed to be. Biotechnology, as Kevin Ulmer had noted, was a going concern, and it was already a species of molecular technology: it worked by making atomic-scale alterations to the DNA molecule, cutting the molecule in half at a selected site and splicing in a short new sequence of atoms. That new sequence of atoms, furthermore, was itself chemically synthesized: it was "synthetic" DNA. By 1978 Genentech, the San Francisco biotech firm, had spliced synthetic human insulin genes into *E. coli* bacteria, which then proceeded to turn out human insulin. Once they got going, the bacteria produced the insulin automatically and without human labor, night and day — exactly as Drexler's molecular assemblers were supposed to do.

Computer technology, meanwhile, had miniaturized electronic components to the point where you could fit a million of them on the surface of a silicon chip a quarter-inch on a side.

Still, neither genetic engineering nor integrated circuitry — nor both of them taken together — added up to nanotechnology. Genetic engineering was limited in its range of possible products to what DNA could be reprogrammed to do and to the substances that its typical host organisms, such as *E. coli,* could be persuaded to make, and those products were biological rather than mechanical. Nanotechnology, by contrast, would manufacture *any* product of *any* type, subject only to the limitations set by the laws of physics and chemistry.

As for computer chips, they had the virtue of being completely general in application — they could store any sort of program or information, biological or not — but they were limited by the fact that a program was merely a set of instructions, that information was merely abstract data. A computer chip by itself didn't do or manufacture anything; it just held stuff in memory. Besides, integrated circuits, small as they were, were not "molecular" devices by any stretch: they were still orders of magnitude too large.

Anyway, it being the logical next step to flesh out the technical details of his molecular assemblers, Drexler instead did something else, spending the next four years, essentially, writing up a popular account of the subject in his book, *Engines of Creation.*

For a dyed-in-the-wool engineer such as himself, this was somewhat puzzling. Why go public with a scheme as wild and woolly as this one before the technical details were even passably well worked out? Why paint vivid word pictures of "the coming era of nanotechnology" before even so much as one paltry designer protein had been coaxed, tricked, or forced into existence? Why not nail down an ironclad scientific case for the whole thing first, and only then proceed to advertise its benefits?

Of course, there were answers. For one thing, Drexler was convinced that he'd already done enough in his *PNAS* piece to motivate a full course of research-and-development work in academia and industry. After all, he'd described what was possible at the molecular level and by what means, and he'd said what some of the benefits were. How could a bunch of forward-looking researchers, seeing all this, not go ahead and actually do it? Science and engineering were already heading downward, toward ultimate control of atoms and molecules, and so the development of nanotechnology, he thought, was all but certain.

The other reason for writing a popular book on the subject was to raise some of the economic and social issues involved. Scientists and engineers, it was commonly observed, did not have an especially good track record when it came to assessing the wider impact of what they'd wrought in the lab. Their attitude seemed to be: "We invent it, you figure out what to do with it."

To Drexler, that was the height of social irresponsibility, particularly where nanotechnology was concerned, because its impacts would be just so broad and sweeping. There were almost too many issues to think about all at once: How would you handle the disruption to the world economy caused by a technology that provided unlimited material goods automatically, at virtually zero labor cost, and with negligible costs for raw materials? What would happen to giant corporations once nanotechnology really got rolling? How did you cope with the mass unemployment promised by the arrival of fully automated molecular manufacturing? What were people going to do with their lives — what were they going to do, period — when physical labor was no longer a necessity, when they could look forward to living for hundreds of years in perfect health and perpetual youth? And so on.

If anything was clear to Eric Drexler, it was that if the human race was to survive the transition to the nanotech era, it would have to do a bit of thinking beforehand. *He'd* have to write the book on this because, all too obviously, nobody else was about to.

But there was yet a third reason for writing *Engines of Creation,* a reason that was, for Drexler, probably the strongest one of all. This was to announce to the world at large that the issue of "limits" had been addressed head-on, that the enemy had been met, engaged, and defeated — or shortly would be. "Limits" really wouldn't matter — they'd hardly exist in any relevant sense — once you'd acquired the capacity to arrange atoms and molecules as you pleased.

So Eric's real message in *Engines of Creation* would be: Look, stop worrying about all those "limits" you've been hearing about. They do exist, sure enough, but they're much farther away than anyone's ever suspected, so far off in space and time, that they're not worth thinking about.

Hang in. Be cool. Don't worry.

Except that there *were* in fact some things to worry about — one or two. There was the tricky little problem of "risks," for example.

Drexler was as terrorized as he'd ever been by the thought of berserker nanomachines. His tiny devices, after all, would have the ability to replicate themselves — indeed, they'd *have* to replicate themselves if they were to have any effect at all in the greater macroscale universe. An individual "assembler" — as he now started calling these molecular manipulators — could line up atoms till kingdom come and hardly make a dent in the Big World. But if the first assembler could make another assembler, and if each of those made yet two more, and so on — much in the way that biological cells divided and replicated — well then, the final result would be a bit different. In fact, matters could get out of hand in very short order.

Suppose the first replicator builds a copy of itself in a thousand seconds. Then, said Drexler, "the two replicators build two more in the next thousand seconds, the four build another four, and the eight build another eight. At the end of ten hours, there are not thirty-six new replicators, but over 68 billion. In less than a day they would weigh a ton; in less than two days, they would outweigh the Earth; in another four hours, they would exceed the mass of the Sun and all the planets combined — if the bottle of chemicals hadn't run dry long before."

That gave you a little different slant on these wonderful, world-saving nanomachines, when you could go from a single self-replicating assembler to the final destruction of the solar system, and all in the space of forty-eight hours.

During one of his visits to the Xanadu house in the summer of 1980, Eric had gotten into a discussion of these out-of-control, self-replicating assemblers with Mark Miller, who immediately saw where it could all lead: they could turn the world into a pile of "gray goo," as he put it.

"The reason gray goo is gray is because it could take over and eat the whole universe, but never become anything interesting."

Soon "the gray-goo problem" was a rather fashionable discussion item at standard gatherings of the nanoclan, whether at the Xanadu house or back at the Space Habitat Study Group in Cambridge. Some of the more freethinking nanotech buffs even went so far as to imagine — if only for a fleeting moment — that the gray-goo threat might actually be held against nanotechnology, that it might even be a reason for suppressing it. Others, including Eric, argued that the better course lay in "containment," figuring out ways of stopping the trouble before it began. And so a whole new slew of strategies was developed.

"One obvious tactic is isolation," said Drexler. "Simple replicators will have no intelligence, and they won't be designed to escape and run wild. Containing them seems no great challenge."

After a while Eric presented this point in a formulation that came to be known as the Tree Sap Answer: "For an industrial replicator designed to operate in a vat of fuel and raw material chemicals, for it to accidentally turn into a replicator that's able to survive in nature . . . that would be about as likely as a car — just 'by accident,' in the garage — being able to wean itself from its diet of gasoline and transmission fluid and go out and live on tree sap in the wild."

Vivid as it was, that was only one answer to the gray-goo threat.

"Better yet, we will be able to design replicators that *can't* escape and run wild," he said. "We can build them with counters (like those in cells) that limit them to a fixed number of replications. We can build them to have requirements for special synthetic 'vitamins,' or for bizarre environments found only in the laboratory."

All of which seemed plausible enough in theory. But, as every hypertext fan knew, there was always another side to the story. There was the famous case of the "HeLa cells," for example, a chestnut item from the history of biology.

HeLa cells were cancer cells taken from a patient named Henrietta Lacks, who'd died at Johns Hopkins Hospital, in Baltimore, on October 4, 1951. Cells had been taken from her body for research purposes and identified by the code "HeLa," the first two letters of each name. The cells were placed in glass tubes, where they continued to divide and replicate in normal fashion, just as if the patient were still alive.

And then those cancer cells got out of hand, running amok and replicating themselves like crazy, constantly, wherever they landed.

Over a period of years, Henrietta Lacks's cancer cells would become the space invaders of tissue culture. So aggressively did they reproduce, in fact, that when, "by accident," a few of them dropped into another cell culture, they promptly took over and wiped out all trace of whatever else had been there originally. Where there had been mouse, hamster, or monkey cells, now there were only HeLa cells. Where there had been cell samples of human larynx, kidney, or heart, now there were only cells from the cancerous tumor of Henrietta Lacks, organisms that had taken over and were now spreading themselves around and proliferating uncontrollably, like a chain reaction. This happened again and again, in cancer labs and medical research facilities all over the world.

The worst of it was, the takeovers occurred before anyone was aware of the fact, with the result that, often as not, researchers had no idea of what was actually in their test tubes, petri dishes, and incubators. Supposedly, the American Type Culture Collection, in Washington, DC, kept on hand a bank of pure and unadulterated cell lines — the nation's "reference cells." But in 1968, seventeen years after the death of Henrietta Lacks, it was discovered that out of thirty-four cell lines in storage there, fully twenty-four of them were in fact HeLa cells.

This was an atrocity, a major scandal. Researchers sent each other HeLagrams. They suffered bouts of HeLaphobia. And who could blame them? — especially after seeing newspaper headlines like "DEAD WOMAN'S CANCER CELLS SPREADING," "GOOF COSTS 20 YEARS OF RESEARCH," and "A SHOCKER FOR SCIENTISTS."

In light of which droll episode, it was plain that Drexler's "containment" strategies for preventing runaway nanoreplicators were not exactly fail-safe. Despite the fact that Henrietta Lacks's cancer cells were not designed to "escape and run wild," HeLa cells did. Despite the fact that human cells started out with built-in "counters" — which is what Drexler proposed installing in his replicating machines to prevent them from multiplying like rabbits — human cancer cells had lost, outwitted, or otherwise bypassed their "counters." And despite the fact that cancer cells normally existed only in the specialized environment of a human body, HeLa cells thrived quite well outside it, in a variety of test tubes, flasks, and other lab glassware — and even in tiny airborne water drops.

"Merely pulling a stopper from a test tube or dispensing liquid from a dropper could launch tiny airborne droplets containing a few HeLa cells," one report said. "When the drops landed on open petri dishes holding live cultures, the HeLa cells began growing so feverishly that in three weeks they overwhelmed the original cultures."

At the height of it all, some Johns Hopkins medical researchers joked that "HeLa, if allowed to grow uninhibited under optimal cultural conditions, would have taken over the world by this time."

If all of this experience proved anything, it was that passive, "containment" strategies might not be enough to stop out-of-control nanomachines. So in addition, Drexler proposed more dynamic methods of dealing with them — specifically, "active shields."

"We can build nanomachines that act somewhat like the white blood cells of the human immune system: devices that can fight not just bacteria and viruses, but dangerous replicators of all sorts. Call an automated defense of this sort an *active shield*, to distinguish it from a fixed wall."

And now there arose this vision of good assemblers and bad assemblers, of one fighting the other, a vast Manichean nanowar of friend and foe, all of it waged at the unseen molecular level.

And suddenly there were two colors of goo, because arrayed out against the gray goo, the enemy, you could have this battalion of nanocops — and cops wore blue — so there'd be blue goo, the defenders, the invisible protectors . . .

Gray goo! Blue goo! Battle lines forming in the Brownian motion! Invisible armies clashing by night!

It would all be funny enough, Drexler thought, if it weren't so serious. It was precisely such invisible battles, waged continuously in the human bloodstream, that kept people alive and well or caused them to die, depending on which way the battle went.

"The gray goo threat makes one thing perfectly clear," he said. "We cannot afford certain kinds of accidents with replicating assemblers. We must not let a single replicating assembler of the wrong kind be loosed on an unprepared world."

* * *

By the spring of 1985, Eric Drexler was on the verge of happening. For one thing, he'd finished his first book, *Engines of Creation,* and his agent, Norman Kurtz, had sold it to Doubleday and Company, which would bring it out the following year. That a major New York house would publish a book calling for the replacement of industrial civilization by a manufacturing system based on countless billions of invisible molecular robots, a book proclaiming the virtues of suspended animation, and dealing with the gray-goo problem in calm, cool, measured tones, was something of a marvel. Nevertheless, there it was.

Then too, in January of 1985 Eric had presented another installment in his seemingly endless series of MIT nanotechnology courses. This one, entitled "Technology and the Limits of the Possible," was the same course, basically, that he'd given time and again over the years, and by now it was turning into a space opera of sorts, with the same plot, theme, music, and scenery, and attended, to a degree, by the same cadre of loyal fans. There was a slight sense of repeating oneself, of preaching to the converted, of this not being a "course" any longer, but rather something on the order of Brother Eric's Nanotech Revival and Molecular Salvation Show.

Some of the more skeptical observers of the growing Eric Drexler phenomenon, indeed, would speak of nanotechnology and its inventor in mock-religious terms. To Kurt Mislow, the Princeton University chemist, Eric was a "prophet" and his followers were "evangelists of the technocratic heaven." To Julius Rebek, the MIT chemist who would later participate in Drexler's Ph.D. oral exam, Drexler and his congregation were "nanotheologists."

Superficially, at least, there were some parallels. By the time he'd finished *Engines of Creation,* nanotechnology was defined by a core set of doctrines: Molecular manipulation was indeed possible. Quantum uncertainty, although real enough, was not a problem for molecular machinery. And though thermal vibration was a problem, it was one that could be overcome by adroit engineering. Proteins could be designed so as to create first-generation molecular machinery, which would then build second-generation machines, which would go on to replicate themselves, and so on and so forth. Atom-by-atom control of matter was both possible and desirable; machines could be made to atomic tolerances; products could be manufactured that were atomically perfect, that had not so much as a single atom out of place — unless it be through radiation damage or other such influences.

More than that, nanotechnology was a distinct way of looking at the world, a weltanschauung or metaphysics. It was such a large and powerful idea that it actually changed the way you thought about nature. Where previously nature had been viewed as recalcitrant, as hostile to human design and intention, now, suddenly, nature was open, welcoming, and controllable. It was as if reality had somehow become permissive, even indulgent.

Nanotechnology promised grand and sweeping changes: an end to aging; a radical postponement of death; conditions of extreme plenty and bountifulness; a freedom from hunger and want. Space travel was around the corner; the stars were within reach.

And it promised all these miracles by virtue of semi-mythological entities, the "assemblers," that not only did not as yet exist, but for whose construction no proven means were available. Not even so much as a blueprint existed.

Nanotechnology promised a big turning point in the future, a millennium — called "the Breakthrough" — when the first working assemblers were actually built. This would be a watershed event in human history, after which things would never be the same again. It was a revolution, moreover, for which people had to start preparing themselves now.

Human life after the Breakthrough would be a state in which brute physical labor had been banished from the planet, all of it having been exported to impalpable molecular machines that would toil on your behalf automatically and without recompense, night and day. As a result there would be a utopian "freedom from work" — a beatific phase that had been last seen Before the Fall. A future of peace, plenty, and enlightenment could be dimly glimpsed through the nanotech mist.

And there was the sense, at MIT — and, increasingly, elsewhere — of being part of the Greater Nanotech Brotherhood, a movement, a fraternity, with leader and followers, united and marching in stepwise fashion along the one true path to molecular salvation. There was a sense of excitement, an awareness of being privy to exclusive, cabalistic, almost forbidden knowledge.

These people, indeed, had the most amazing secrets inside of their heads. They knew what could happen through nanotechnology, they knew where it could lead. More, they were present at the creation. They were a formative part of it all. They were the founding mothers and fathers of a great and potentially extremely dangerous new order of things. There was a veiled feeling of being one of the Elect, the Select, the Knowledgeable, the Chosen.

Frequently, too, Drexler inspired a sense of awe in the listener, just as if he were the charismatic leader Brother Eric — not that he ever pretended to be anything of the sort, and in fact he detested any variety of "Trekkie" cultishness. Still, he was a fabulous success at whipping up the crowds with his visions of molecular manufacturing, of lightsailing away to the asteroid belt, of rocketing off toward the Virgo cluster, and all the rest. He spoke of skyscrapers that would need aircraft warning lights not at the top but at the base; halfway up the structure, which would be a thousand kilometers high, you'd have to worry about colliding with satellites.

Anyway, after giving the January course, Drexler's next major MIT event was on March 14, at a meeting of the Students for the Exploration and Development of Space, SEDS. This is where Dave Forrest, an MIT grad student in metallurgical engineering first heard Brother Eric preach. And what an awe-inspiring experience it was!

"I was definitely not prepared for what he said," Forrest remembered.

Drexler was up there in tie and jacket, holding forth on various schemes for dropping the cost of earth-to-orbit transport by two or three orders of magnitude. To Dave Forrest, who had ambitions of becoming a NASA mission specialist (a.k.a. shuttle astronaut), this was astounding. To Drexler, though, such schemes were as nothing. They, after all, were the old technology.

"This scenario," Drexler had said, "completely ignores the possibility of radical technological advances sometime around the years 2000 to 2010."

"And he proceeded to describe what he thought those advances would be," Forrest recalled. "Breakthroughs in computational speed, robotics, and this thing called nanotechnology. All accelerated through computer-integrated manufacturing which shortens the time to design and manufacture a part.

"Then he got to the part about nanotechnology. My notes faded out here. He showed slides of DNA and ribosomes in rapid succession. What did me in, though, was the drawing of the T-4 bacteriophage. It was a virus, but it looked like an alien robot. A real, self-replicating molecular machine.

"He talked about the potential consequences of nanotechnology. Complete, utter destruction. The possibility of a totalitarian regime taking control of the entire planet. Or a utopia beyond what I had imagined possible. Perfect health. Intelligent furniture. 3D flat-screen TV. Superstrong materials made out of diamond. And all in my own lifetime.

"I'm an optimist so it was really wonderful to hear all of this. I felt confident that we humans could get through the dangers he had outlined. And here he was virtually assuring me that I would get into space — it was just what I wanted to hear.

"I spoke with him briefly afterwards about the materials-engineering possibilities. It was the only thing I was competent enough in to talk to him about. And in that area I judged that he definitely knew what he was talking about. So I signed up on a mailing list being carried around by a woman named Chris. I remember thinking, as I walked away from the meeting down near-empty halls, that I felt like I had been propelled five hundred years into the future."

Well! With Brother Eric up at the pulpit and Sister Chris working the crowd, was this not truly the Religion of the Molecules? Could anyone be blamed for seeing a churchly element in the whole thing? For a religion, it seemed to have everything but choir practice.

All that said, there were also one or two wee differences between nanotechnology and the old-time religion. For one thing, instead of being based on faith, nanotechnology was based on reason: on science, technology, and engineering. Its patron saints were named Watson and Crick. For another, there were no sacred texts involved; instead, there were equations, statistics, graphs, data. If this was a religion, it would be the first one in world history where the closest approach to a Bible was the *CRC Handbook of Chemistry and Physics.* Instead of the Gospel according to Saint Matthew there was the Born-Oppenheimer approximation. In place of the Revelations there was *The Molecular Biology of the Gene.* And where Proverbs used to be there were now these crisp little technical disquisitions on subjects like thermoelastic damping and phonon viscosity — whatever they were.

The nanotech millennium would be ushered in not by prayer but by experimental advances in protein folding. And the wonders it promised, miraculous as they were, would be brought to pass not by magic or miracles or spirits from the beyond, but rather by machines. Tiny machines, invisible machines — but still just a batch of essentially prosaic and blah *appliances.*

And, for all the nano miracle-equivalents that awaited off in the distance, there were an equal number of terrors lurking off on the same horizon. Many a nanoid would retire at night with visions of molecular utopia dancing in his or her head — only to wake up the next morning in the grip of a gray-goo nightmare, or worse.

Nanotechnology was not an unalloyed blessing, that was for sure. And, appearances aside, it was not quite the religion its critics made it out to be.

About a week after the March 14 SEDS meeting, all who'd attended got a letter in the mail from Chris Peterson informing them that they were now members of the Nanotechnology Study Group, the NSG.

> Dear NSG Member:
>
> Eric Drexler's Nanotechnology talk for the MIT SEDS (Students for the Exploration and Development of Space) went well. He emphasized space applications, and touched on applications in computers, defense, and medicine. SEDS videotaped the talk, and all attendees signed up for the

NSG mailing list. They are welcome at all activities listed here.

The "activities" in question included, first, a March 27 discussion meeting in room 773 of the AI Lab; second, a party.

SATURDAY, MARCH 30 8:00 PARTY

Come meet NSG/SEDS members and friends in a relaxed atmosphere at Chris and Eric's: 518 Putnam Ave. Apt. 9, Cambridge. Feel free to help with refreshments if you like; just bring anything plausible. If you'd like to bring more than one non-NSG/SEDS friend, please call Chris first.

Third, and most intriguing of all:

APRIL 6-7 WEEKEND RETREAT

Get out of the city and come discuss nanotechnology and our future at 'Camelot' (one of the MIT Outing Club's lodges) hidden away in the woods at the foot of New Hampshire's White Mountains. Bring sleeping bags, hiking boots, and $3/night for lodging and $7 for food (includes two breakfasts, two lunches, and Saturday dinner).

And finally, there was this last cryptic announcement:

APRIL LECTURE

A general lecture by Eric Drexler, directed at the MIT community, is being planned. This must be done in April as Eric is leaving the area in May.

Leaving the area?

The fact was, Chris and Eric were moving to California. There were just these last few events before the great departure: a meeting, a lecture, a party — and the weekend retreat. That would be Eric's leave-taking, apparently.

In fact it would be more than that. It would be Eric's East Coast summing-up, his final analysis, plus his own indirect and subtle way of telling his disciples: "Go ye therefore into the world, and preach the nanotech gospel to every creature."

* * *

Along about six o'clock Friday night, April 4, Dave Forrest was standing in the main lobby at MIT — his sleeping bag, gallon of water, flashlight, and clothes on the floor beside him — waiting for his ride to the White Mountains. A few minutes after six, Eric showed up, and then so did Scott Jones, an NSG member who worked over at the AI Lab. All of them walked down the steps to the street, where Chris was waiting in her aging Chevy.

"I don't remember too much about the ride up," Forrest said, "except that Eric did most of the talking, and a lot of the talk was between Scott and Eric about computers. I didn't understand very much of it. I do remember Eric doing lots of mental calculations. At one point he lost track after doing about five or six side-calculations, to get numbers for his main calculation. Chris chided him, 'Why don't you use my calculator, that's what it's for!' But he didn't listen, and finished his calculation mentally."

Three hours later, having driven past the cabin four times in dense fog, shining flashlights at people's houses for the street numbers, the crew finally located the trailhead to the cabin. Another car drove up an hour or so later, bringing NSG members Bruce Mackenzie, Stewart Cobb, Dave Blackwell, and Dave Lindbergh. The last two, Kevin Nelson and Christopher Fry, arrived next morning. That made ten: Chris, Eric, and the other eight.

The Eight Apostles.

A more straight-laced group was not imaginable, this bunch of science, engineering, and computer types in flannel shirts and blue jeans, all come here to the White Mountains, "to talk about," as Eric put it, "what needs to be done."

One thing that needed to be done at the weekend retreat was to bring everyone involved to the so-called Miller point. This was of course a reference to Mark Miller of the Xanadu Project, who, early on in his thinking about nanotechnology, had come to a rather stunning realization, namely that nanotechnology would let you do almost anything. And he meant that quite literally: Eric's tiny programmed robots could manufacture virtually any object that was allowed by nature.

Once people had access to such a technology, he realized, the basic conditions of human life would be transformed forever after. People would no longer have to work for a living, they could do whatever they wanted. They could get everything they needed, or could possibly desire, for free, from this race of tiny machines. They could live practically forever, and so on and so forth. Miller was a big science-fiction fan, on speaking terms with every last utopian vision, and now, suddenly, all of it would be possible, and maybe even within his own lifetime.

The full meaning and import of all this came to him late one night in a surge, in a rush — "I'm not sure if it was three or four in the morning, but it was sometime in the middle of the night," he said — it came to him that after the breakthrough, things were going to be quite different. Really different.

Eric Drexler, Chris Peterson (front, third and second from right) and members of the MIT Nanotechnology Study Group at the Weekend Retreat in New Hampshire's White Mountains, April 7, 1985 (*David R. Forrest*)

He had to tell Eric. Eric, he thought, had not quite understood the true scope and dimensions of his own creation. The two of them had talked for hours on end about how strange things would be in the future — but not *this* strange.

"I called him up almost in a panic. I said, 'You know, things are going to be really different!' And he said, 'I know, I've been trying to tell you that.' And I said, 'No, no, I mean *really* different!'

"And then as we talked I realized that he had already understood. He'd been trying to explain it to me but he was being a little incremental about it, trying to kind of drag me along. And then I'd taken the final few steps myself."

Taking those final few steps — realizing that nanotechnology would change absolutely everything, that once it got going nothing would ever be the same again — this soon came to be known among the nano faithful as "reaching the Miller point."

At the weekend retreat, Eric had a simple plan for bringing all those present to the Miller point. After hours of discussion of assemblers, replicators, nanocomputers, and everything else, after breaking for lunch, during which the talk quite naturally turned to Eric's "food machine," which would rearrange common waste materials into a steak without ever killing a cow, and to dishes that would clean themselves up and return themselves back to the cabinet — just the ordinary household miracles — some time after all this, Eric produced a Magic Marker and a big sheet of paper, display-board size.

He asked people to name all the different technologies they could possibly think of; he listed these on one side of the sheet — Column A. He then asked which of those technologies would be unaffected, completely untouched, by nanotechnology. Any which fell into that category would be placed on the other side of the sheet — Column B.

But no one could think of any technology that wouldn't be changed by the arrival of the nanotech turning point, and so nothing got put in Column B. Eric then asked: Well, then, what might remain unaffected in the *non*-technological realm?

"How about love?" someone asked. Well, but consider the effects of understanding, in molecular detail, the biochemistry of "falling in love." And they all immediately nodded their heads. Love was not placed in Column B.

"How about religion?" someone else asked. Well, but truly advanced artificially intelligent systems, which nanotechnology would help create, they'd raise some religious questions, wouldn't they? Were artificial minds souls? Were they spirits? And so on. Religion was not put in Column B, either. After an hour or so of this, Column B remained mostly empty, except for one or two items on the order of "stud poker," "square dancing," and the like.

And at length everyone could see how truly different, how completely new, things were going to be, After the Breakthrough.

"This is something that is going to fundamentally change the world," Dave Lindbergh remembered thinking. "And we all sort of felt like we were in a privileged position because we were one of the few who knew about it, and we had the chance to think about it."

"A lot of hair standing up on the back of necks, goose bumps on forearms and so forth," Kevin Nelson recalled.

And now it was time for Eric to issue his charge, his mandate, his "Go ye therefore . . ."

"What I did was, I briefly summarized some of the things we had discussed," he recalled much later. "I said, 'Here's what's happened; here's what we seem to think needs to happen; and here's some of the things that are appropriate for us to consider, in terms of action.'

"And then I said that very shortly I was going to be leaving for California."

And at that precise moment he pulled on his parka and gloves, went out the door, and hiked off toward the lake.

part ii

Escape

8

Tiny *Tale* Gets Grand

In 1983, after almost a quarter of a century, Richard Feynman gave a new "Room at the Bottom" talk, this at the Jet Propulsion Laboratory, at Caltech.

"This is a kind of 'There's Plenty of Room at the Bottom' revisited," he said at the start. "I've been asked a number of times to reconsider all the things that I talked about twenty-three years ago, to see how things have changed, and see what the new situation is."

The new situation, basically, was that what had seemed unbelievable in 1959 was now old hat. Computers, which were hulking giants back then, taking up whole rooms, had been reduced to the point of near-invisibility. Feynman projected on the screen a slide of a computer chip.

"This was incredible twenty-three years ago," he said, "but it's where we are today. We can reduce information to a very small scale."

He told them about the tiny motor that he'd offered a prize for, just a sixty-fourth of an inch on a side. That had been built within a few months — faster than even he had thought possible.

"I wanted to get a motor that couldn't be made directly by hand," he said, "and I proposed a sixty-fourth of an inch. And at the end of the talk, Don Glaser, the Nobel Prize winner in physics, said, 'You should have said a *two-hundredth* of an inch on a side, because a sixty-fourth on a side is just about possible by hand.'

"I didn't believe Glaser," Feynman added. "But anyway, the guy made it by hand!"

Feynman had the motor with him, Bill McLellan's motor, right then and there. It was mounted in a little portable display case that he now started circulating around the audience. The box was equipped with a crank that you could turn to make the motor run. It was also equipped with a magnifying glass, so that you could actually see it run.

"Try it without the glass first," he said. "As you can see, you can't see it."

A wave of laughter as the engineers one by one took a look at this incomprehensible speck.

Tiny motor!

"I believe that with today's technology," Feynman continued, "we can easily — I say *easily* — construct motors one-fortieth of that size in each dimension, sixty-four thousand times smaller than that, than McLellan's motor, and that we can make thousands of them at a time, all separately controllable so that you can turn on one or the other."

But why?

"What use would such a thing be? Now it gets embarrassing. . . . There is *no* use for them, and I don't understand why I'm fascinated by the question of making small machines with movable and controllable parts. I'm fascinated, I don't know why. And every once in a while I try to find a use . . . and I know there's been already a lot of laughter in the audience, but just save it for the uses that I'm gonna suggest for some of these devices."

You could spread out the tiny motors on a flat surface and attach their drive shafts to a series of shutters, so that you could control the passage of light through an array of tiny holes. Such an arrangement would let you project patterns of light onto a screen — just like television.

Not that this was a "use." We already had television.

"I don't think projecting television pictures has any use anyhow, except to sell more television pictures, or something like that. I don't consider that a 'use,' advertising toilet paper."

His second use was to build self-cleaning surfaces: "If you had little rollers on a surface, then if dirt falls on the surface you keep it clean all the time by having the rollers . . ."

More laughter.

"I feel very embarrassed because I haven't really thought of anything good to say," Feynman said. "But here I am, having to say something."

It became clear, finally, that although he loved these tiny machines, that he was crazy about them, he nevertheless hadn't figured out the least little thing they'd actually be good for. Not a single practical application.

Well, games, maybe.

"The purpose is no doubt entertainment, like video games. This machine that you can control from the outside has a sword, okay, and it gets in the water with a paramecium . . ."

Wild laughter. And so it went for about an hour or so.

"I would love to be led on in a direction where I could see the technical applications, so I could see the economic steps that would lead us to be making smaller machines, and then smaller machines," Feynman said, toward the end. "And if I were really successful at this talk, I would have outlined the economic sequence of applications. But my uses don't seem to be useful enough. Perhaps someone here will see that one of them isn't so dumb, and that it is worthwhile to start moving in that direction. But at the present time, I don't know any way.

"I keep getting frustrated in thinking about these machines," he said. "I want somebody to think of a good use, so that the future will really have these things in it."

In May of 1985, about a month after the New Hampshire retreat, Chris and Eric left for the West. Others of the clan were moving out there, too: Mark Miller, Roger Gregory, the whole Xanadu team.

"The reason we all came to California," Miller explained later, "is that Roger Gregory and Chip Morningstar were in Michigan, I was in Texas, Drexler was in Massachusetts, and Phil Salin was in California. The whole group of us really wanted to be together in one place because we were clearly this intellectual community, and we were running up tremendous long-distance bills talking to each other. We needed to rendezvous and form our own community, and California was the only place we would all agree to move."

Chris and Eric had started out towing a trailer with the full complement of their worldly belongings. But the car started making the worst groaning, grunting, and grinding noises they ever heard, before they'd even left Boston, whereas their itinerary included crossing the Rockies.

"This is a bad idea," Chris said after a few minutes. So they stopped at a moving company, transferred their extra stuff to a van, and headed off again in a car still loaded with books, clothes, houseplants, the manuscript for *Engines of Creation*, and assorted computer gear. They had a Zenith Z89 computer, a separate Heathkit machine, plus some Stone Age 85K disk drives that Marvin Minsky had "loaned" to Eric.

"At one point I realized that no, he never was going to want these back," Eric said.

To Drexler, never much one for traveling, the trip was mostly a blur.

"I seem to recall going through Gary, Indiana. We went through Nebraska. A side jaunt down to visit my uncle, Andy Gassmann, in Castle Rock, Colorado." Then they were in Utah.

"In Utah we encountered amazing rain. Thunderstorms and hailstorms. In fact there were big piles of hailstones by the side of the road: they'd bounce off the sides of the road cuts and pile up into heaps. It was really quite something. And then in some of the salt-flat areas in western Utah — it had been raining there, too — there was a thin layer of rainwater on top of the salt. And because the water was so shallow, what you had was this glassy smooth water surface on top of a salt flat and the clearest and crispest reflections of mountains I could ever recall having seen."

And finally, California. The shining light! The research institutes! The computer chips!

By the time they crossed over the border, the White Mountains weekend retreat was already a month in the past. The meeting, Drexler thought, could hardly have gone better. Just as he'd intended, the nanotech faithful had realized — once he'd gone out the door and walked off — that the problem was now in *their* hands.

They'd sat there in stunned silence for a moment, nonplussed, dumbfounded, like those big stone heads on Easter Island.

"Jesus Christ," one of them said. "This is terrible."

"Yeah, it's like now *we're* responsible," said another. "We're the ones who know about this. And we're the only ones, so it's up to us to take care of it, right? I mean, we've gotta do something."

They had to tell the world about the coming changes. Warn them. Prepare people, for better or worse.

At length they settled on a battle plan. They'd continue the Nanotechnology Study Group that Chris and Eric had begun, they'd make it into a regular fixture at MIT. They'd hold monthly meetings — maybe even biweekly. They'd sponsor talks, hold conferences and retreats, publish a newsletter. They'd make MIT into a flaming hotbed of nanotech activity.

Except for development work, that is. Nobody ever said a word about that. Development, they thought, was something that would go of itself. Nanotechnology was inevitable, unstoppable. You wouldn't have to make it happen; it was more likely that you couldn't prevent it from happening.

Drexler came back from his walk in the woods an hour or so later. The group members told him that they'd grokked his message completely and had even started to make some plans.

Which Eric very much liked the sound of. It was just exactly what he'd been hoping for, that these people would take it upon themselves to spread the word, to do what they could to get his ideas taken seriously by people at the highest levels of science and technology — who were not at all difficult to find at MIT.

So he gave his disciples a round of applause. They cleaned the place up, collected their gear, and left.

It was a wee bit incongruous, of course, this small assembly of nanotech kids out for a weekend picnic, charting the world's destiny.

"There *was* some sense of incongruousness," Dave Lindbergh recalled, much later. "There was a certain sense of, here is this group of a dozen guys in the backwoods of New Hampshire coming to plot the future of the universe. And yet there was also a sense of, hey, if not us, who? We're the ones who knew about it. We're the ones who had a vision of what was going to be happening. We knew something that everybody else didn't."

"It was like stumbling across a toxic-waste dump in the middle of town," Kevin Nelson said. "You're going to tell somebody. The arguments seemed simple enough and clear enough so that we had no question that nanotechnology was going to work, and no question that if it worked it would cut both ways, it would be a two-edged sword. So to us it was a clear and present danger."

"We knew that if we weren't careful we could destroy the world in a really weird way," said Christopher Fry, another of the participants. Fry had worn his Bullwinkle cap for much of the weekend — his pro-wildlife, anti-hunting gesture.

Not everyone was equally quaking in their boots, however. "I remember during the ride home, Chris and Eric asked me if all this didn't scare me," Dave Forrest said. "It didn't. Maybe because I used to read science-fiction stories, this didn't seem so unfamiliar. In a sense, I'd already seen scenarios for the destruction of the earth, or for utopian futures. What I wasn't ready for was the possibility that it would happen in my lifetime."

But it would if Eric Drexler had anything to say about it. Making it happen in his own lifetime would become Drexler's personal project while out in California.

While they looked for a place to live, Chris and Eric lived with two of their friends, Phil Salin and Gayle Pergamit, who were in the private-rocket business (or were trying to be: private-launch businesses had extremely short half-lives). Drexler and Peterson soon moved into a two-bedroom garden apartment at 86 Renato Court in Redwood City. It was close to Xanadu headquarters in Palo Alto, and also to Stanford, where Eric had been appointed "Visiting Scholar" in the computer science department. Not one of your top-ranked positions at Stanford: just a formal university affiliation plus library privileges. No duties, no boss, no salary.

For a while, he and Chris would live off savings. Every so often Eric would do some consulting work for Xanadu, maybe give some paying lectures. That, they hoped, would be money enough to get by.

Finally, and at long last, Eric would start writing his technical text.

Just as Drexler was settling into his visiting-scholar post at Stanford, another prime nano-event was taking place across campus. Tom Newman, a Stanford grad student, was about to try for the second Feynman prize.

Newman was knee-deep in doctoral-dissertation research when Ken Polasko, a buddy in the electrical-engineering department, came in with a copy of Feynman's 1959 talk. Polasko pointed out the passage where Feynman had offered "a prize of $1,000 to the first guy who can take the information on the page of a book and put it on an area 1/25,000 smaller in linear scale in such a manner that it can be read by an electron microscope."

"We should be able to do that," Polasko told Newman. After all, both of them worked with electron microscopes and electron beams every day of the week. "We've got everything we need right here in the lab. It ought to be easy."

Tom Newman wasn't too sure about that. The lettering, after all, would be extremely tiny: it would have to be because it was the scale at which the whole *Encyclopaedia Britannica,* all twenty-four volumes worth, could be written on the head of a pin. Even with all the experience he'd had at doing things at the microscale, Tom Newman didn't know if such a thing was possible.

"My initial reaction to the Feynman challenge was, it would be *too* challenging," he recalled. "I looked at what it would take to write letters versus writing simple patterns like lines and dots, and the size-scale of the letters was quite fine, quite small. And so I wasn't too interested, at least at first."

On the other hand Tom Newman was also a whiz at electronics. He had been ever since age eight or so, which was when his father, an electrical engineer for Martin-Marietta, had introduced him to the world of solder guns, capacitors, and the smell of hot electrical circuitry. Later, at the University of Florida at Gainesville, Newman majored in electrical engineering, graduating with a 3.98 grade point average.

"I got a B in 'Marching Band,' " he recalled. "I missed practice."

After that it was off to Stanford to learn microelectronics, integrated-circuit design, and electron-beam technology. Electron beams, it turned out, were the workhorses of the very small scale. Not only were they were used in electron microscopes, to see the tiniest objects, they were also used for fabrication purposes.

"You can scan that beam anywhere you want on a workpiece," he said, "and turn the beam on and off at high speed, using computer control. So you take basically a scanning electron microscope, or the equivalent, add some auxiliary equipment, and what you have is an electron-beam fabrication tool."

Such a tool could be used to make integrated circuits, in a process known as electron-beam lithography. Newman had once designed an integrated circuit with 25,000 transistors on it.

"That was considered to be a fairly substantial number back then," he said. "Today a million transistors on a chip is not a big deal."

By the time Ken Polasko showed him the Feynman piece, both of them were probing the limits of electron-beam lithography and had almost reached the level of quantum effects. Anyway, as skeptical as he was about going for the Feynman prize, Newman jotted down a few notes to himself and put them off at one corner of his desk.

Every morning he'd come in and see those notes. They were a distraction; they seemed to be sending special messages out to him. After a while they were burning a hole in his head.

"Well, how hard could it be?" he'd think. After all, he was quite accustomed to working at the tiniest imaginable scales, and he was even pretty good at it: when his 25,000-transistor chip had actually been fabricated, it had worked perfectly the first time out. So just for the hell of it he calculated how small a page of text would have to be in order to be reduced 25,000 times.

Pretty small, he learned to his shock. In fact, the page in question would have to be reduced to roughly the size of *one* of the transistors on a computer chip. The individual letters would wind up being only about fifty atoms across.

Fifty atoms!

But in 1985, that was a doable feat. Basically, all you'd need was a computer program that caused the electron beam to scan back and forth in patterns forming the twenty-six letters of the alphabet.

"It looked like it *would* be possible, so I started kind of making a strategy. One problem was, I was near the end of my thesis, and it was to the point where you realize you've really got to get working to be out of school in a reasonable time. And my adviser was encouraging me to avoid superfluous projects. So I felt a little bit guilty about diverting any attention to this, but the way it worked out was, one weekend my adviser was away for three days — he went to Washington, DC, or somewhere, for a panel — and basically I could work by myself in the lab without being observed. I decided that I would work on it for three days, and see what I came up with."

At about that time, Newman sent Feynman a telegram asking if his 1959 prize offer was still open — the one about writing the tiny text.

After just a couple of days, Newman had made progress.

"This was basically a programming challenge," he said. "Most programming challenges, if you know they're realizable and you have some knowledge of the language, you can do it by staying up late and drinking coffee."

Which he ended up doing. He now had a program that would control the electron beam as it wrote out the first page of the Charles Dickens novel *A Tale of Two Cities.* It was a favorite book of his, and besides, the opening words were so well known — "It was the best of times, it was the worst of times" — that when it came to deciphering the output, anyone fluent in English would be helped along by memory.

Then one day he got a phone call from Feynman.

"He called me right in the lab," said Newman. "It was a little strange because I was just sitting there working and I got a phone call and I picked up the phone, and it was Richard Feynman.

Which was fairly stunning. I didn't know what to say, or how to handle myself, really, just being a grad student. But he said, 'What are you guys *doing* up there?'

"Anyway, we talked briefly about it, and he encouraged me to go for it. And he sounded pretty excited about it himself."

Soon Newman had his tiny *Tale* written out at the proper size. It took the electron-beam apparatus only about sixty seconds to write out the full page of text.

"That's quick," said Newman. "It would take five hundred minutes to write the whole *Tale of Two Cities,* if it were five hundred pages long. That's not absurd, not out of reach. But then it was discouraging to realize how long it would take to write the whole Library of Congress."

The main technical difficulty of writing that small, he discovered, was physically locating the text again once he'd actually written it on the surface. At the scale of 1/25,000, after all, a pinhead was an immense area.

"Finding the page of text turned out to be a challenge because it was so small, compared to the area we were writing on. When you're at low magnification it's hard to see things in the electron microscope, but if you zoom in you're looking too close, and it takes forever to look around. So we needed to make a little road map of each sample: there's a speck of dirt here, a little chip here, and we'd use that to home in on it. But then once you saw it on the screen of the microscope, it was fairly legible."

In early November 1985, Newman put a package together and sent it off to Feynman by Federal Express. It contained a print of the first page of *A Tale of Two Cities,* plus supporting evidence that the whole page had been reduced to the required dimensions.

Two weeks later, Feynman put a check in the mail.

"Congratulations to you and your colleagues," he wrote. "You have certainly satisfied my idea of what I wanted to give a prize for. Others have apparently made as small or smaller marks, but no one tried to print an entire page. And on a 512×512 dot printer! Each dot is only about 60 atoms on a side. I can't quite imagine the square 1/160 mm on a side onto which all that is printed. It would be 20 times too small on a side to see with the naked eye. Only ten wavelengths of light. The entire *Encyclopaedia Britannica*, perhaps 50,000 to 100,000 pages of your size would be on less than 2 mm on a side — the head of a small plain pin. . . . Can application to computers be far behind?

"As promised long ago, I am enclosing a check for $1,000 for your accomplishment."

And that was the end of it . . . except for one or two curious inquiries in the wake of all the publicity. A publisher of Buddhist texts wrote in to ask Newman if he'd have any interest in creating a line of microscopic prayer books.

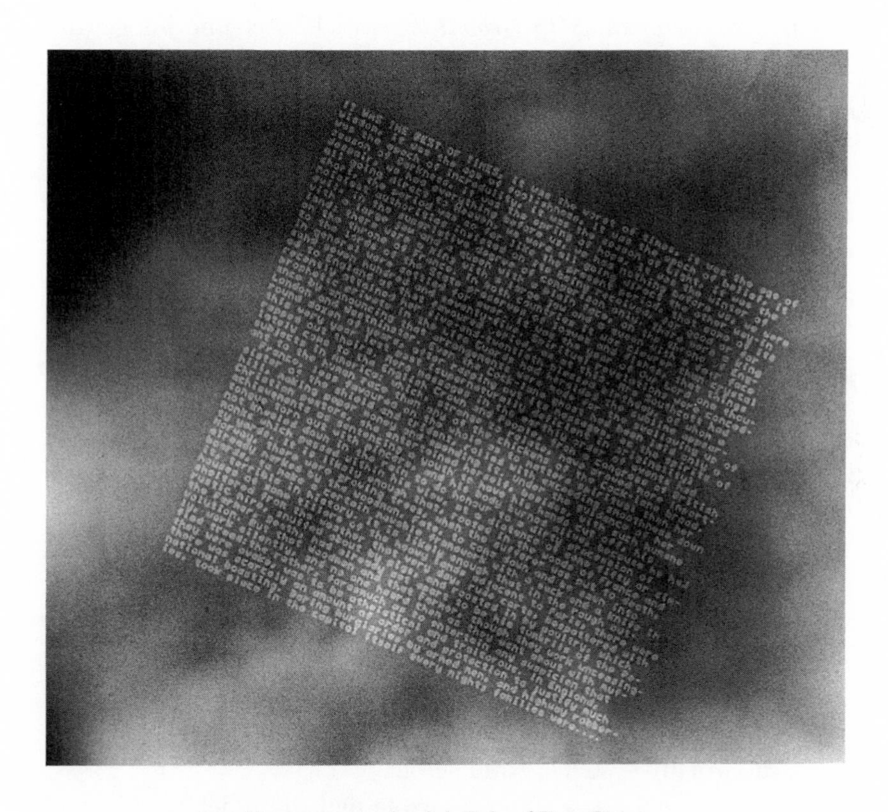

First paragraph of *A Tale of Two Cities*,
written by electron beam at 1/25,000 scale reduction;
awarded the second Feynman prize, 1985. (*Stanford University*)

"Miniature texts have potential ritual uses that are very intriguing," the letter said. "One requirement would be that the texts be in Tibetan script, rather than the standard Roman alphabet, but we already have or could modify a bit map of the Tibetan alphabet."

But Tom Newman would write no more tiny texts. He'd gotten his doctorate and had taken a job at the IBM Thomas J. Watson Research Center, in Yorktown Heights, New York, where he'd try to make some tiny computers.

To say that *Engines of Creation* was an unusual book was not quite to appreciate the reality of the situation. It was Drexler's personal answer to *Limits to Growth.* It was his scheme for getting through the next few hundred years or so in perfect health and perpetual youth, all the time awash in conditions of unheard-of material abundance. And it was his analysis of the economic and social issues posed by the great changeover. For many, reading the book would be a transformative experience: you seemed to inhabit a different universe afterward.

At the bottom of it all, of course, were Drexler's tiny molecular machines, the "engines of creation" of the title. These invisible programmed robots would do everything that needed to be done. Some of them would do manufacturing work, making everything from paper clips to lunar rovers. Others ("engines of healing") would specialize in biological tasks, repairing any and all ailing cells. Others ("disassemblers") would do cleanup jobs, breaking down toxic wastes into harmless or even helpful substances.

Unfortunately there was also the possibility that still other engines ("engines of destruction") would bring with them the gray-goo threat, the prospect of "destroying the biosphere," annihilating the planet, and other choice treats. But the book included an entire chapterful of schemes for defeating aggressive replicators, for making "trustworthy systems," and so on.

Mainly, *Engines of Creation* explained the inner workings of nanotechnology. Current technology, Drexler argued, was really a vestige of the ancient past. The traditional methods of shaping matter hadn't much changed in thousands of years, and all of them involved pushing around vast quantities of raw materials forcibly, by coercion.

"Thirty thousand years ago, chipping flint was high technology," he wrote. Later on people manipulated matter by heating, rolling, and pounding things into submission, but people still did that today: "We now cook up pure ceramics and stronger steels, but we still shape them by pounding, chipping, and so forth."

That was Drexler's picture of "modern manufacturing"— pounding. You had this vision of convicts out there in the prison yard swinging sledgehammers under the hot sun, pulverizing rocks, clubbing them to bits . . . chips flying everywhere, clouds of smoke rising up, the ground shaking underfoot. It was loud, noisy, messy as hell.

Steam shovels! Smelters! Smokestacks! Bessemer converters! Oil cans! Grease cups! Diesel fumes! Gas, dust, heat, waste!

And what did such ugly, brute-force methods get you in the end but a flawed and imperfect product, anyway. There'd be these tiny little cracks here and there: hairline flaws, weak spots, faults, defects. You never even knew they were there until the part gave way, which always occurred at the worst possible moment, as when the engine-mount bolt on the jumbo-jet failed during takeoff.

Drexler didn't like any part of that stuff. His own sympathies — his aesthetic sensibilities, if truth be told — lay with the light, the fine, the almost-weightless. His lightsail, after all, was to be manufactured by a spray process — you'd spray it into existence — producing a sheet of foil that would stretch for miles in all directions. The sail would be fully functional — it would do actual work — notwithstanding that it was just a few molecules thick. As for more conventional objects — ordinary three-dimensional basics such as a rocket engine, for example — well, his recipe for that was one of the dramatic high points of *Engines of Creation.*

Conventional rocket engines, Drexler explained, were normally made by casting, milling, whittling pieces of metal into shape; crucial components were welded, bolted, riveted together. The end product was a gross and dull artifact marred by seams that were visible and by defects that weren't. A nanotech rocket engine, by contrast, would be put together molecule by molecule, and the final product would be as precisely ordered and as perfect as a crystal. In fact it would actually *be* a crystal, because a nanotech rocket engine would not be fashioned out of anything so coarse and lowly as aluminum or titanium. It would be made, instead, out of diamond and sapphire. And it would be built in silence and utter calm.

The fabrication took place inside of a sealed vat, into which various chemicals were pumped.

"To begin the process, the operator swings back the top of the vat and lowers into it a base plate on which the engine will be built. The top is then resealed. At the touch of a button, pumps flood the chamber with a thick, milky fluid which submerges the plate. . . . This fluid flows from another vat in which replicating assemblers have been raised and then reprogrammed by making them copy and spread a new instruction tape (a bit like infecting a bacteria with a virus)."

The assemblers then "grow" the engine from a programmed "seed." They physically add diamond molecule to diamond molecule, building up the engine according to its atomic blueprints, like molecular bricklayers building a wall.

"Finally, the vat drains, a spray rinses the engine, the lid lifts, and the finished engine is hoisted out to dry. Its creation has required less than a day and almost no human attention. . . . Rather than being a massive piece of welded and bolted metal, it is a seamless thing, gemlike."

Much of it is empty space, honeycombed with strategically placed voids. "Compared to a modern metal engine," Drexler wrote, "this advanced engine has over 90 percent less mass." The empty spaces, "patterned in arrays about a wavelength of light apart, have a side effect: like the pits on a laser disk they diffract light, producing a varied iridescence like that of a fire opal."

Seamless. Light. Diamond and sapphire.

That was a Drexler rocket engine.

E*ngines of Creation* did not get a smash reception. Not exactly. There were only two major reviews, one of them in the *New*

York Times, which described the book as a "clearly written, hopeful forecast, remarkable for an unembarrassed faith in progress through technology."

(*Faith?* Drexler had a hard time with that one. He had more fear of nanotechnology than he had "faith" in it.)

The reviewer doubted that the nanotech revolution would happen anytime soon. "If biology is any measure," he wrote, "it will be a long time before scientists get nanotechnology humming. Consider that, while the earliest living cells probably had proteins much like some of our own, only after about two billion years of evolutionary trial and error did such nanomachines gather themselves into anything as complex as a nerve."

(Evolution? Two billion years? This guy's assuming that engineers are no brighter than cosmic rays!)

The only other full-scale notice appeared in *Technology Review,* an MIT publication. Written by Hans Moravec, a computer scientist at Carnegie Mellon University, it pretty much embraced the whole nanotech vision.

"I find his central point convincing," Moravec wrote. "Atomic scale construction is not just possible but inevitable in the foreseeable future. The human benefits from this technology will be without limit. Self-replicating machinery will be able to create a mind-boggling abundance for all."

Provided, of course, that the gray-goo problem could be solved — about which Moravec was not so confident.

"Genius gray goo might infiltrate the guards with its own agents and gradually subvert the defense from inside, for example. Rogues could also escape beyond the frontier of controlled space and mass for a frontal assault."

The worst of it was, we'd be facing these problems sooner rather than later.

"Drexler estimates that the time to these developments is under fifty years. I concur. Our accelerating technology will soon reach a kind of escape velocity that will carry us into a new and radically different world."

* * *

Feynman gave his "Room at the Bottom, Revisited" talk just one other time. This was at Esalen in 1984.

Here was the hot-tub and massage crowd come to hear this world-class, Nobel Prize physicist, barefoot and in shorts for the occasion. "Tiny Machines," the talk was called — not that anybody got the drift.

"What's this 'Tiny Machines' really gonna be about, anyway?" they asked him.

It was about small machinery, he said: "I'd say, 'You know, very small machines.' But it doesn't work. I am talking about *very small machines*, okay?"

The Esalen version covered what was, by now, the standard territory. In the Q&A period, though, someone finally asked The Question.

"How'd you ever come up with all this?"

It went back to the early 1950s, Feynman said.

"I was playing on the beach one day. I began to think of how small a space I'd need to write a book. And I thought the way to do it . . . in those days an easy way would be to evaporate layers, one after the other, on a big area. And that's easy to do, five atoms for one bit, and so forth. I was absolutely amazed that a book could be put on a wire two inches long. I went around to all the people on the beach, and said, 'Do you know it's possible?' "

No one cared. "I didn't sell any bananas."

That was as much as Richard Feynman ever said about the origins of "Room at the Bottom." But that was not the whole story. In fact it was not even the main story, because it only dealt with information storage, which was only half the picture, and not the most important half.

The important half had to do with molecular manipulation, the idea of being able to "arrange atoms the way we want; the very *atoms*, all the way down!" It had to do with machines that built tinier machines. Where had *that* stuff come from?

That — irony of ironies — had come from *science fiction!*

In the fall of 1958, when Caltech's Jet Propulsion Laboratory was just getting into the lunar and planetary exploration business, Feynman's good friend Al Hibbs had come up with an idea for exploring the moon and the planets.

"I thought, 'Why not send a robot to the moon?' " Hibbs recalled. "A robot operated by a man, like the 'Waldoes' in Heinlein's story."

In 1942 Robert A. Heinlein, the science-fiction writer, published a novelette called *Waldo*. In the story, an inventor named Waldo develops a system of mechanical hands — "grotesquely humanoid gadgets" — that were remotely controlled by the hands of a human operator. Any motion the human hands made, the mechanical slave hands duplicated on a smaller scale. Naturally, the slave hands were put to the task of creating even tinier hands, until, at the very bottom, "the last stage was tiny metal blossoms hardly an eighth of an inch across."

Al Hibbs recalled the Heinlein story and decided that this remote-controlled, master-slave business was just the ticket for getting around the moon cheaply. "I was so enamored of this notion — of a Waldo-like robot operated by a man — that I wrote up a patent application."

Dated February 8, 1959, the year of Feynman's talk, the patent application described a miniature, remote-controlled mechanical man.

"This declaration applies to a man-machine combination with application wherever human types of activities must be carried out in a hostile environment," it said. "The machine part of the combination is a scale model of all or part of a human being, with the joints activated by servo-controlled actuators, and sensory organs (electrical or mechanical) to provide in electrical form information on (1) joint position of all moving joints, (2) optical (television) coverage from one or two optical sensors (located in the relative position of human eyes), (3) sound (if necessary), (4) pressure on important areas (such as the bottom of the feet, palms of hands, and finger tips), (5) temperature (if important), and (6) other sensory features (such as attitude with reference to gravity)."

The declaration went on like this for five pages and included a sketch of the wee mechanical man, plus a detail of its robotic fingers and their complex of joints, sensors, and actuators.

"It was in this period, December of 1958 to January of 1959, that I talked it over with Feynman," said Hibbs. "Our conversations went beyond my 'remote manipulator' into the notion of making things smaller — half-size machinist robots working with half-size machine tools, making robots and tools a half-size smaller, and so forth. I suggested a minute surgeon robot, armed with a sword-sized scalpel, injected into the bloodstream, and going about whacking off plaque, cancer cells, and so on.

"He was delighted with the notion," Hibbs said, "and we talked for a while about the limitations, such as the wavelength of radiation for the optical sensors, sizes of lenses, resolution, and so on. Not then, but some time later — weeks? months? — he told me his idea about information: that this was not limited by the kind of radiation wavelengths we had been talking about, but could be stored and manipulated in something very small, like electrons, where the limits were established by the jitters of thermodynamics.

"He thought this might be a good idea. I agreed. He wrote up the talk."

As it was, Hibbs never filed his patent application.

And Feynman never found any use for his "tiny machines."

"Maybe someday they'll find a use," he said at Esalen. "Give or take twenty-five or thirty years, there'll be some practical use for this. What it is, I don't know."

9

Astrid and Priscilla

Ever since the dawn of quantum mechanics in the 1920s, one thing that everybody "knew" about atoms and elementary particles was that working with them was out of the question. Isolating a single atom absolutely could not be done. In fact, even entertaining the thought of atoms as clearly defined entities with exact locations was a sign of retrograde thinking, a barbarity, a relic from the olde-tyme Newtonian physics. The very notion reeked of ruff collars and powdered wigs; it was the quantum-mechanical equivalent of a quaint superstition. In modern times everyone "knew" that elementary particles existed in a realm of quantum indeterminacy and inherent fuzziness, and that only statistical collections of them were even conceivable.

"The individual particle is not a well-defined permanent entity of detectable identity or sameness," said Erwin Schrödinger, the German physicist. "We *never* experiment with just *one* electron or atom or (small) molecule. In thought experiments we sometimes assume that we do; this invariably entails ridiculous consequences."

"Is it right to speak of a 'particular' carbon atom?" asked Primo Levi, the Italian chemist. "For the chemist there exist some doubts."

Despite all that, teachers from kindergarten through graduate school somehow felt completely justified in representing elementary particles as dots on the blackboard, as if atoms, nuclei, electrons, and all the rest were standard Newtonian objects, just like back in the good old days. Millions of kids stared at those isolated blackboard dots, and none of them ever saw a problem.

Except for Hans Dehmelt.

"I mean, there's a mild discrepancy, isn't there?" Dehmelt said. "My teacher drew a dot on the blackboard declaring, 'Here is an electron.' But in a parallel lecture in beginning quantum mechanics, I learned that an electron at rest, or for that matter any particle at rest, wouldn't be localized at all."

That "mild discrepancy" stayed with him for the rest of his days. In fact, it bothered him to the point that he spent much of the rest of his life trying to do what everyone "knew" could not be done: isolate an individual particle and make it stand still.

Eventually, he succeeded.

Dehmelt, it's true, was a born experimentalist. "I started out doing various experiments as a kid in my mother's kitchen," he recalled. "Mild explosions and things, the usual things one does in one's mother's kitchen."

Mild explosions?

"I remember one thing in particular — this was later — I decided to invent a simple way to remove my very few first whiskers, with an evil-smelling solution, a depilatory. It was a complete failure. It just burned my skin."

Dehmelt was more successful in physics proper, enrolling at the University of Göttingen in the 1940s. That was where his professor, Richard Becker, put that little white dot on the blackboard, saying it was an electron.

Just a little white dot! But to Dehmelt, that dot was a provocation, a challenge. It bugged him.

He began to wonder why you couldn't in fact isolate a real electron and make it stand still like that tiny white blackboard dot. What was to stop you?

Maybe you could do it with magnets, he thought, or with charged plates: you could build a little glass cage of some type, run wires around it, and control the electrical currents in such a way as to keep an electron inside it suspended in space. Not bunches of electrons or even just a few. Rather, a single, solitary, and lone electron.

Words and phrases popped into his head. *An electron at rest in space. An atom floating forever at rest.* They ran through his cortex like radio waves, those phrases. Finally he hit upon the best one of all: *A single atomic particle forever floating at rest in free space.* He even made up an acronym: SAPARIS, a Single Atomic Particle At Rest In Space.

Later, as professor of physics at the University of Washington, in Seattle, Dehmelt worked at confining electrons in ion traps, which were magnetized plates inside cathode-ray tubes that kept stray particles from causing spots on the screen. After a while he enjoyed some modest success at this, immobilizing small clouds of electrons for a fleeting second or two, much as if they were a swarm of bees. He experimented with related devices, "Penning traps," which enclosed smaller groups of electrons inside electrically generated fields of force.

Finally, in 1973, Dehmelt was able to get most of the unwanted electrons out of the trap, letting them out little by little, like air from a balloon, until there were only a few remaining.

And then there was only one, a single electron. It buzzed back and forth inside the Penning trap like a fly in a bottle. A very small fly, this "monoelectron oscillator," as he called it, vibrated back and forth with amplitudes that were almost too tiny to measure. Still, it was oscillating by this slight amount, and so he had not quite achieved his goal. When he published articles about his captured electron, therefore, he spoke of a "single electron *almost* at rest in free space."

Dehmelt kept that first electron cooped up for only a few days, but soon he was keeping single electrons penned up for weeks, then even months at a time. Finally he kept one caged for ten months — almost a year! — before it collided with the side of the bottle and vanished.

Then, in 1980 Dehmelt trapped a single atom inside of a complex system of lasers. He kept it there for just a matter of days, but it was long enough for him to grow fond of his little trapped atom.

"Why not give it a name?" he thought.

"The well-defined identity of this elementary particle is something fundamentally new, which deserves to be recognized by being given a name, just as pets are given names of persons," he said in the otherwise sane pages of *Science.*

This would be Astrid. Astrid Atom.

Later, in 1984 — the year in which Feynman gave his Esalen talk in which he said, "Never mind making computers with wires, make them with atoms . . . you just put the atoms down where you want them," and also the year in which Drexler was in the middle of writing *Engines of Creation,* in which he said, "Assemblers will let us place atoms in almost any reasonable arrangement" — Dehmelt trapped another type of particle, a positron (a particle with the mass of an electron, but with a plus instead of a minus charge), and kept it caged for three months.

This was Priscilla. Priscilla Positron.

"Here, right now, in a little cylindrical domain, about 30 μm [microns] in diameter and 60 μm long, in the center of our Penning trap resides positron (or anti-electron) Priscilla, who has been giving spontaneous and command performances of her quantum jump ballets for the last three months," he said in the journal *Atomic Physics.*

Astrid Atom and Priscilla Positron — Hans Dehmelt's pet particles.

Having now done many times over what everyone "knew" was impossible, Dehmelt became the sworn enemy of what he regarded as "a still persisting wave of quantum-mystification in the literature," the notion that atoms, electrons, and suchlike, were somehow not fully "real" objects. They were real enough, he decided, if you could isolate them for months at a time, give them names, and make them perform "quantum jump ballets."

Indeed, Dehmelt seemed to take a special glee in making the quantum theorists of yore turn over in their graves. His experiments, he once wrote, "laid to rest Wolfgang Pauli's assertions — backed by Niels Bohr — that the spin magnetic moment of the electron could never be measured." It *could* be measured, because he'd measured it. Ha!

As for Schrödinger's claim that "we *never* experiment with just *one* electron or atom or (small) molecule," Dehmelt conveyed what he thought of that proposition by giving one of his papers the in-your-face title "Experiments on the Structure of an Individual Elementary Particle."

And when, in 1988, he published a paper giving a new value for the electron radius, he was finally able to use his all-time favorite heavenly phrase as the title: "A Single Atomic Particle Forever Floating at Rest in Free Space."

The very next year, Dehmelt won the Nobel Prize for physics.

Anyone who doubted Drexler's contention that atoms were as real as marbles first had to reckon with Astrid and Priscilla. Not to mention Hans Dehmelt.

One of the first things Chris and Eric did after establishing themselves in California was to start a new organization out there, a sort of nanotech-West. This was the Foresight Institute, a nonprofit venture that would publicize, explain, promote, and, they hoped, guide the development of nanotechnology. For nano fans it was world headquarters.

Eric had by this time pretty much come around to the view that the safest thing to do about his nanotech dream was to advertise and advocate it as much as he could. He had not always felt that way. Earlier, in stage one of his thinking, he'd thought it best to keep quiet about the prospect of molecular manipulation and manufacturing; that way, people would be less likely to actually develop it and, whether by malice or accident, convert the known world into the famous gray goo.

That was his attitude, anyway, from the time he came up with the idea in the spring of 1977 until the time reality intervened in the form of the *Semiconductor International* piece, describing protein-based electronic circuitry, that Chris had brought home for him. That's when he learned that some of this stuff was apparently on the verge of happening without his so much as uttering a peep. But if nanotechnology was going to be invented anyway — by somebody or other, someplace, whether he published his ideas about it or not — then at least he ought to do what he could to minimize its various threats and dangers, which most people, apparently, were not even aware of. But the question was, how best to minimize them?

The issue was moderately complicated and called for a major bout of tactical planning. Plainly, you couldn't let matters take their own course: that way lay the gray-goo disaster, irrational political upheavals, attempts to stifle nanotechnology before it even began, or God only knew what other sorts of random and nonlinear human behavior. Nanotechnology would have to be managed, and therefore you had to develop a game plan, an overall public attitude or stance to take toward it, a strategic methodological position. Position number one (public silence) had already been preempted by the course of events, so it was necessary to come up with an alternative.

Position number two favored active publicity.

As the New Hampshire weekend retreat had shown, nanotechnology would change just about everything, and virtually overnight. But you couldn't spring such a drastic and sudden change upon people without telling them about it beforehand. They'd go nuts at the thought of tiny invisible machines overrunning their homes and offices, putting them out of work, invading their very bodies — "I don't want a bunch of itty-bitty robots flying around in *my* backyard!" and so on. It would be future shock in the extreme.

So you'd have to explain it all very carefully to them ahead of time, telling people what the benefits were, showing them how the various pitfalls could be identified and avoided, so on and so forth. That meant a massive, worldwide educational campaign. ("Bunches of itty-bitty robots have been there all the time, you know. They're called bacteria.")

Position number two, however, still fell short of supporting, advocating, or actually promoting nanotechnology. In fact, during the time Drexler was maintaining position number two he even went so far as to *deny* that he was endorsing nanotechnology, or that he'd ever done so.

"I have not *advocated* nanotechnology," he said, "I have advocated *understanding* it. Reporters, hearing me describe a technology that can accomplish many long-sought goals, often assume I must think it is an unalloyed blessing, or at least a good thing. My position seems just a shade too subtle to fit a simple, stereotyped story: I believe that in our diverse, competitive world, basic human motivations make nanotechnology effectively inevitable, and that, in light of this, we need to understand its great potential for good and ill so that we can formulate and act in accord with effective policies."

Well, maybe.

But shortly after that followed position number three, which favored active support and development, and the sooner the better.

His reasoning here was that if nanotechnology was going be developed anyway, whether he helped it along or not, then it was crucial that it be invented here in America, or at least by one of the free democracies wherever they were located, East or West. This was crucial because the first nation to develop nanotechnology would thereby become the world's dominant power, "the Leading Force."

That nation, whoever it was, could build weapons that no other country would have defenses against. Its citizens would become healthy, wealthy, and young overnight. It would be Them against everyone else.

Moreover, it was not out of bounds to imagine one of the more unspoiled worldly monarchies being the first to develop nanotechnology. Nanotechnology research, after all, was not "big science" in the usual sense. You didn't need anything like a Manhattan Project or an Apollo program or a Superconducting Supercollider effort to get the thing going. Conceivably, you could do it in a garage. You could do simulations of molecular machines on a personal computer; you could create billions of molecular structures in a test tube; you could custom-make DNA in a desktop synthesizer. All you needed for the great breakthrough was a laboratory, some extremely smart people and programming, and lots of luck at getting things right.

So Eric's strategic methodological position number three was that he not only should publicize nanotechnology, he should actually hurry it along as much as he could. The Foresight Institute, therefore, would do its part to hasten the day.

The institute, at first, was little more than a post office box and a telephone answering machine in Chris and Eric's kitchen. Mostly, people heard about it from reading *Engines of Creation*, which ended with the words:

> If you want to keep in touch with these developments, and with efforts to understand and influence them, please get a paper and pen and send your name and address to:
>
> The Foresight Institute
> P.O. Box 61058
> Palo Alto, Calif. 94306.

Within a few months of its publication, *Engines* had pulled in a few hundred members.

Finally it was time to do some actual design work. Thus far, nanotechnology had been little more than an exercise in fantasy. Nowhere in his various talks had Richard Feynman reduced any of his "tiny machines" to numbers or equations, much less actual blueprints. But then neither had Eric Drexler. *Engines of Creation* was unadulterated prose: there was not an illustration, formula, design, or equation (except for $E = mc^2$) anywhere in its almost three hundred pages. Even his technical *PNAS* piece "Molecular Engineering" contained nary an equation or sample molecular design. For a revolution in advanced technology, this would be the first one ever brought off by sheer verbalization.

So there was this slight matter of design work to attend to. For Drexler, this would mean providing a bunch of nitty-gritty specifics and concrete details. He'd have to say what his molecular machines would look like and how they'd function. He'd have to figure out what the actual component parts of his machines would be, where they'd go, what physical orientations they'd be arranged in, and how they'd be joined together. He'd have to determine how many individual atoms each separate part would be composed of, and precisely which chemical elements would be used to construct them. He'd have to calculate the physical stresses and strains involved, deal with the thermal energy (kT) issue, think about the rules of chemical valence, and on and on. Most important of all, he'd have to describe these things not merely qualitatively, in words, but quantitatively, with numbers.

Of course, there were a few challenges involved.

For one thing, none of it had ever been tried before. This was a completely new field, pure and virgin territory. But however romantic it might have sounded to be "pioneering uncharted realms," the reality of it was that there was a total lack of preexisting borders, landmarks, and milestones to work from. Engineers in the usual fields had at their disposal endless ready-reference texts of all sorts: there were manuals, handbooks, guides, dictionaries, encyclopedias, directories, charts, graphs, tables — all of which summarized decades if not centuries of engineering experience and practical lore. A bridge builder, for example, had within easy reach all kinds of data about tensile strengths, bending moments, elastic limits, elongation factors, coefficients of linear expansion, of girders, cables, rods, and everything else, ad infinitum.

But nothing like that existed if what you were designing was a molecular machine, a mechanical gadget built out of a countable number of atoms. Except for Feynman, von Hippel, and a few other wild-eyed dreamers, few had seriously considered the atomic realm to be a proper locus for engineering work — or even a possible locus — and so there were no "molecular-engineering data" available as such. Rather there was the Great Vacancy.

Where, for example, did the budding young molecular engineer go to look up such facts as how much physical bending could be withstood by a rod consisting of a hundred lined-up carbon atoms? By what distance (in nanometers) would such a rod stretch under a given tensile load (in nanonewtons)? Or would it break in half immediately instead of stretching? By what amount could you twist such a rod before it sheared apart? How much thermal vibration would such a rod undergo across a given range of temperatures?

How fast could a molecular robot arm of a given length, diameter, and mass be made to swing back and forth through a given arc? How fast could a molecular wheel be spun before it broke up and flew off in all directions?

If you didn't know the numbers then you couldn't do a plausible design. But you couldn't find those numbers in a molecular-engineering text, because there was no such thing anywhere in existence.

Answers to such questions could be had, of course, but you'd have to supply them yourself, piecing them together, taking one value from a chemistry text, another from a physics book, another from a reference work on quantum mechanics. In other cases the answers would have to be derived, calculated, or extrapolated out from known values. All of which was complicated by the fact that in most cases you couldn't simply extrapolate directly from the macroworld to the nanoscale. You could extrapolate down from four feet to two feet, or even to two inches, but not to two atoms.

You'd have to apply correction factors, because not all physical qualities scaled up and down in the same ratio. When you took a given object and varied one of its physical properties, the others didn't necessarily vary at the same rate. The volume occupied by a cube, for example, didn't increase or decrease at the same rate as a given linear dimension did. A cube 2 inches on a side occupied a volume of $2^3 = 8$ cubic inches, but if its linear dimensions were doubled to 4 inches on a side, its volume would not merely be doubled (to 16 cubic inches), it would jump to $4^3 = 64$ cubic inches.

These scaling factors often had the most unexpected real-world consequences. One might intuitively think, for example, that if you doubled the size of a given airplane it would fly just as well. But in fact it wouldn't, because while the plane's wing area, and therefore its lift, varied as the square, its volume, mass, and weight each varied as the cube. The aircraft's weight would therefore increase much more rapidly than the lift of its wings, which meant that sooner or later it would fall out of the sky like a rock.

What all this scaling business meant for Eric Drexler was that when he took some macroscopic object — a gear or bearing, for example — and designed a nanoscale version of it, he had to take due account of all the relevant physical phenomena. He couldn't simply shrink a given part down to atomic size and expect it to function like its Big World counterpart.

The foremost difference between the macro- and nanoworlds was that some of the phenomena that showed up at the nanoscale didn't even exist in the Big World. Chief among them, of course, was the graininess of matter, the fact that at the molecular level all of the available building materials came in the form of discrete individual chunks, the atoms.

This had both advantages and drawbacks. The main advantages were that atoms were prefabricated, perfect, and existed in almost infinite supply.

"Atoms need not be made," said Drexler. "They are both flawless and available without need for manufacture."

Each atom of a given element, moreover, was absolutely identical to the next, meaning that they were perfect building blocks for anyone seeking to design precisely patterned arrays of them.

"These patterns will be either entirely correct or clearly wrong. In stacking part on part there will be no buildup of small errors, as there is in conventional systems. Nanomechanisms won't wear out. So long as all the atoms in a nanomechanism are present, properly bonded, and not in a distinct, excited state, the mechanism is perfect."

On the other hand, there were drawbacks, or at least complications, to designing atomically perfect machinery — complications that had no Big World analogue.

For one thing, there was no cut-to-fit with atoms as there was up in the Big World. In carpentry, for example, you could cut a piece of wood to almost any arbitrary size, and you could do the same with metal, plastic, or any other normal building material. But you couldn't do it with atoms. There was no such thing as sawing a carbon atom in half so that it would fit in perfectly just where you needed it. Nor could you compress it down, squash it in, or force it into a shape much different from the one it had already. The atoms of a given element came in just the one discrete size and shape they came in, no bigger and no smaller, and that was how you had to take them.

But there was a second atomic peculiarity that had no Big

World equivalent — the fact that atoms were also subject to *forces* that didn't exist at the macroscale level.

Molecules attracted and repelled one another almost as if they were living creatures with likes and dislikes. Very few entities in the Big World acted this way, the main exception being magnets. But in the small world, atoms and molecules regularly attracted and repelled each other as a function of distance: at short distances they repelled (this was known as "overlap repulsion," an effect of the "overlap repulsion force"), while at longer distances they attracted each other (due to the so-called "van der Waals forces").

These and other molecular influences could work for you or against you. At any rate, they were factors that an atomic-level engineer had to take into account, contend with, and adjust for.

Designing for the world of atoms, then, would be no mean feat. It was like designing for the Big World, only with a bunch of extra added physical idiosyncrasies, oddball forces, and nanoscale quirks thrown in. "Design constraints," which is what these were, were nothing new in the engineering world: an engineer always had to work within a set of limitations imposed by materials, costs, energy requirements, and so on.

Molecular design constraints, however, were off in a class by themselves.

First on Drexler's list was a bearing.

Later, some critics would claim that Eric was forever purveying "science fiction" because he was trying to build starships even before the Wright brothers flew.

"You don't build a Gothic cathedral until you have the vaulted arch," said Vince Rotello, an MIT chemist. "Drexler doesn't have the vaulted arch, but he's out there saying, 'If we had these warp drives we could conquer the universe.' "

The fact was, however, that Drexler had started out by designing a bearing, which was about as low on the totem pole as you could possibly get; only the simple machines of antiquity — such as the lever, screw, and inclined plane — could be regarded as being more basic. A bearing was a mechanical device that allowed motion to take place between two parts of a machine.

"A fundamental question in making mechanical systems is how do you make moving parts," said Drexler. "And moving parts require bearings, at least in the usual implementations, so bearings are a natural place to focus."

A natural place particularly in view of the fact that making bearings out of atoms did not seem possible. Atoms were inherently knobbed and bumpy, and so were the force fields surrounding them. How was it possible to make a smoothly rotating bearing when both the shaft itself and the cylindrical sleeve surrounding it consisted of knobbed and bumpy atoms?

"It's not obvious that you can have two bumpy surfaces that slide over one another smoothly," said Drexler.

Nor could you turn to lubrication for help. Feynman had addressed that issue, although to little effect.

"Lubrication involves some interesting points," he'd said in "Room at the Bottom." "The effective viscosity of oil would be higher and higher in proportion as we went down. If we don't increase the speed so much, and change from oil to kerosene or some other fluid, the problem is not so bad."

But in Drexler's view kerosene would be of absolutely no help at the atomic level: "From the perspective of a typical nanomachine, a kerosene molecule is an object, not a lubricant."

In his 1981 *PNAS* paper Drexler had suggested using sigma bonds — a type of chemical bond between two atoms — as bearings. "Sigma bonds that have low steric hindrance can serve as rotary bearings able to support $\approx 10^{-9}$N," he'd said. "A line of sigma bonds can serve as a hinge."

Somewhere in the land of nanomachines, though, you'd have a need for bearings that were bigger and stronger than sigma bonds were. You'd need an axle that rotated within a housing, for example. How was it possible to make such a thing out of bumpy atoms and still have the axle rotate smoothly?

Drexler started out with the simplest conceptual situation, a small inner ring of atoms rotating inside of a slightly larger outer ring. In order for the two circles of atoms to slide past each other freely, the first order of business was to make sure that the atoms in the one ring didn't accidentally bond with those of the other: any such bonding and the two rings would be stuck together like glue.

But you could prevent such bonds from forming by using atoms that had no free bonding sites available. Such atoms were "saturated," meaning that all of their open bonding sites were already taken up by other atoms. Drexler imagined making the rings out of fluorinated diamond molecules: on such molecules all bonding sites were already occupied by fluorine atoms, and no accidental bonding could possibly occur.

As for the bumpiness problem, two concentric atomic surfaces appeared to be the world's unlikeliest candidates for smooth rotation. It was like putting a small pearl bracelet inside of another slightly larger bracelet, and then forcing the two to rotate past each other while touching. Any motion that existed would be lurchy and jolting, absolutely the worst prospect for a universe of atomically perfect machines.

But matters were not all that bad down there among the atoms. First of all, atoms were not as hard and unyielding as pearls: their outer surfaces were electron clouds, making them somewhat soft and springy. Second, the atoms making up the two rings would not be in actual physical contact: the so-called "overlap repulsive force" quite conveniently prevented that from happening. The force was an invisible cushion between molecules, a protective barrier that kept each one at arm's length from the next. This peculiarity of the atomic realm actually worked in favor of atomically precise bearings: rings of atoms would slide past each other not as hard knobs but as soft cushions.

Still, they were rounded cushions, and so there would still be some residual lurchiness left in the motion.

There was an artful way around this, too, however, as Drexler soon realized. If you arranged the bumps strategically so that the hills of the outer ring would be prevented from dropping into the valleys of the inner ring, then there would be little if any up-and-down component to the rotation. You could effect such an arrangement by adjusting the number of bumps on the two rings, making it physically impossible for them to fall into each other's open spaces.

"For example, imagine that the outer ring had twelve bumps and the inner one had six," said Drexler. "Then each of the inner bumps could fall between two bumps on the outer ring, and motion between the two rings would be very bumpy. But imagine that the outer ring had *eleven* atoms. Then there is no orientation of the inner ring that lets its bumps simultaneously fall into a lot of different holes on the outer ring. All orientations turn out to be so nearly equivalent that you get a smooth, nonbumpy rotation as a consequence."

There was a simple way of guaranteeing that the bumps on the two rings would automatically mismatch: if the number of bumps on the two rings shared no common numerical divisors, then they couldn't line up so that the bumps of one ring fell into the valleys of the other.

By this point, then, Drexler had discovered a design principle for an entire class of atomically precise bearings. He called them "van der Waals bearings" (later, "overlap repulsion bearings") because they were based on the short-distance repulsive forces between molecules. Such bearings ought to work perfectly without any lubrication, he thought, because the two atomic surfaces, held apart by overlap repulsion, would never actually touch.

Eric now wrote up a technical paper explaining his van der Waals bearings along with some other molecular designs — for screw bearings, linear bearings, journal bearings, as well as for molecular gears, drive shafts, and chains. Almost certainly it was the first such paper of its kind, which led to the problem of where to publish it.

"What journal do you submit it to?" he said later on. "It's not in any existing field, so what journal would take it? What journal would referee it? What would they do with it? It could be perfectly sound work and yet not fit into the existing system of specialties and disciplines and journals. Furthermore, a lot of the interesting work to be done assumes as background a set of synthetic capabilities that doesn't exist yet. I was doing theoretical studies of what you could build if you had positional control of chemical synthesis. Well, do you preface that paper with an explanation of positional synthesis, so that now the rest of it has some context? If you did, it would now be a paper on two different things."

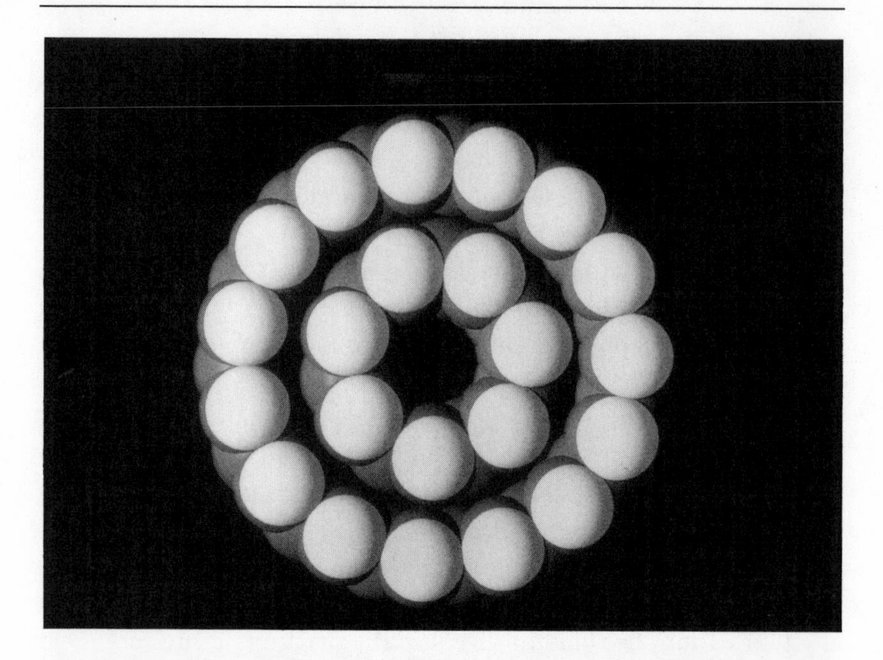

A molecular bearing. Since the number of atoms on the outer and inner rings have no common numerical divisors, the rings can slide past each other without bumpiness. (*K. Eric Drexler and Ralph C. Merkle*)

This would always be a problem for Eric Drexler, one that he'd never solve to his own satisfaction, much less that of his critics, who'd complain that "he doesn't publish in the right journals." It marked the beginning of the paradigm-shift problem that he'd encounter again and again, the hostility of the old to the new, of the entrenched viewpoint to this radically different nanotechnological order of things. So much of his work crossed traditional disciplinary boundaries, ran counter to long-held assumptions, and upset conventional wisdom that editors of the reigning mainstream journals often didn't know what to make of it. Sometimes they responded by doing nothing at all. He once submitted a paper, "Engineering Macromolecular Objects," to the journal *Protein Engineering* — this was in 1989. Two and a half years later, in 1992, he still hadn't heard a word from the editor, and so he withdrew it. (He published it as part of his book *Nanosystems.*)

In the case of his molecular gear and bearing designs, Drexler had been invited to a microrobotics conference on Cape Cod to be held in the fall of 1987, and he gave his gears and bearings paper there.

"It drew a positive but puzzled reaction," he recalled much later. "The attitude seemed to be: What's the point of trying to understand things we can't yet make?"

Which would become a common question. The paper, "Nanomachinery: Atomically Precise Gears and Bearings," published in the conference proceedings called *IEEE Micro Robots and Teleoperators Workshop*, could not be said to have a big readership. But those who did read the paper often had that same reaction.

Sheldon Glashow, a Nobel Prize–winning physicist at Harvard, read the paper and said: "There's nothing kooky about it, except for the premise that you can actually make these things. You make a wild premise and then worry about the little details, fine. The man wants to worry about axles and whatnot. But the issue is to *make* this small critter that does these things.

"I have my doubts about this game since it hasn't really begun yet," he added. "I mean, how are we going to do this? Biologically? Are we going to work our way down? That's not clear."

George M. Whitesides, chairman of Harvard's chemistry

department, also looked at Drexler's designs for molecular gears and bearings.

"The toughest issue is how to *make* them," he remarked. "To rely on an 'assembler' is not satisfactory; it simply gives a name to something you don't know how to do. Much of this type of discussion is like discussing automobile fabrication in a world in which there are no lathes, drill presses, welders, stamping presses, and so on. The tools of production usually determine what is makable, long before thermodynamic limits set in. To set those aside avoids the really hard questions.

"The answer to the question 'Would these devices work if you could build them?' is 'Maybe.' "

Others raised some narrow technical points. Oxford University chemist P. W. Atkins, for instance, also read Drexler's paper.

"There are a number of ideas in it that I would question," he said. "The structures he mentions are plausible, so I have no quarrel with that. However, I am very skeptical about whether they would work in the manner suggested. The whole thrust of Drexler's argument is based on the existence of van der Waals (nonbonding) interactions. However, under the stresses anticipated (although I must confess I find it difficult to relate the calculated forces to quantities that are relevant to chemistry), van der Waals forces will give way to covalence."

("Covalence" referred to chemical bonds formed by the sharing of electrons. Such bonds were much stronger than van der Waals forces: it was the difference between Krazy Glue and Post-it Notes.)

"Take the mention of a chlorine atom acting as a single bearing," Atkins said. "It is only a naive chemist who would regard Cl [chlorine] as monovalent. One is certainly taught that in high school, but in the afterlife one encounters Cl atoms that form more than one covalent bond. Almost certainly, some of the structures he mentions would effectively weld together as electron redistributions gave rise to covalent bond formation."

But Drexler, as he tended to, had an answer to each objection.

"It is indeed 'only a naive chemist who would regard Cl as [always] monovalent,' because in some molecules it does indeed form more than one bond," said Drexler. "But in the proposed structure a particular chlorine atom would be bonded to a single carbon atom, and hence would be monovalent.

"The possibility that 'structures would effectively weld together as electron redistributions gave rise to covalent bond formation' is a fatal problem for some structures and a non-problem for others. The difference is a matter of choosing the right design, one in which a chemical rearrangement that would form the unwanted bonds would cost too much energy to happen at any appreciable rate."

And so on and so forth, point for point, ending up with his universally accommodating, all-purpose offer: "Any design that is found to have a specific chemical instability will cheerfully be replaced with a different device that avoids the problem but serves the same function. Engineering design projects can afford to make a few mistakes, provided they can be fixed later."

In the end, even P. W. Atkins conceded that Drexler's designs might just possibly work out as planned. "This criticism does not mean that the devices will not work," he said. "One would certainly like them to! My attitude is that it would be wonderful if he were right, and he is being very imaginative, but I need to be convinced."

Anyway, Eric Drexler now had his van der Waals bearing.

Not quite the warp drive.

10
Monotony, Hate, and Utopia

Although according to popular legend MIT engineers were officially uninterested in subjects like the social consequences of a given technology — or the social consequences of anything, for that matter — Drexler had been motivated from the start by the idea of practical benefits, by the thought that his molecular-engineering scheme would help people.

"He's always had a focus on improving the human condition," Chris Peterson said. "I mean, that's why he was into space colonization and that's why he's into nanotechnology. He has a focus on human beings and how they are, and their welfare and their health. And it's not just human beings as opposed to the rest of the biosphere that he cares about; he's intensely interested in living things, and that includes animals and plants. He's kind of a shy person and so he's not a bubbly, effusive, 'I love everybody' type, but he really does care."

Superficially, anyway, the major social boon brought to us by nanotechnology would be freedom from physical labor, this as a result of the world's work being turned over to the assemblers. No longer would humanity have to toil for its daily bread. No more getting up in the morning, going in to the office, having to be friendly with people. No more time clocks. Clearly, the promised land had been sighted, and it lay not all that far ahead.

Or at least that was the first-glance view of the situation. But, of course, there being a cloud for every silver lining, deeper analysis led to some darker suspicions. Supposing for the sake of argument that assemblers could in fact, with trivial exceptions, do the work of the world, what then was going to be left for people to do? How would they fill their endless days, especially across thousands of years? When it was nine o'clock on Monday morning and you didn't have to be at the office, what then did you actually do?

And particularly what did you do for money? No more time clocks, after all, also meant no more paychecks. What would happen, after the revolution, to those great and valued social institutions known as the market system, the gross national product, the world economy?

Indeed, the more you thought about it the more it appeared that nanotechnology was a two-edged sword not only in terms of its gray-goo physical risks, but also because of the dangers it posed to the social and economic order. Under the worst-case interpretation, nanotechnology was the ultimate Luddite nightmare: people would be displaced by machines once and for all — machines, furthermore, that were much too small to wreck. How could you smash the machines (as the Luddites always wanted to do) when you couldn't even see the damn things?

Such questions turned up, naturally enough, as rather amusing discussion items at more than one meeting of the MIT Nanotechnology Study Group. Whatever else could be said of the NSG membership, one complaint you couldn't make was that they ducked the hard issues. Just the reverse: if there was a potential problem with nanotechnology, the NSG wanted to know about it. They wanted to solve it, if possible, or least get it discussed. That was their oath from the weekend retreat, and it was one that they were duty-bound to honor.

The NSG had started a "Nanotechnology Notebook," a collection of articles on various nano topics, and some of the pieces in it addressed the unemployment issue. Obviously, Drexler's invention was not the first one to threaten mass unemployment: the Luddites themselves went back to the early 1800s, and in more recent times the prospect of automation, robotics, and artificial intelligence revived the old controversy. But if the papers in the Nanotechnology Notebook were any indication, there was no agreement on the subject even among the experts. Basically there were two viewpoints: one was that automation did not mean job loss; the other was that it did.

James Albus, a robotics researcher, took the first view. "There is not a fixed amount of work," he said. "More work can always be created. Work is easy to create. There is always more work to be done than people to do it." Artificial intelligence would itself create jobs. "Building the automatic factories is a Herculean task that will provide employment to millions of workers for several generations." And so on.

Nils Nilsson, an artificial-intelligence researcher, took the precise opposite position. "Even if AI does create more work," he said, "this work can also be performed by AI devices without necessarily implying more jobs for humans." People *will* in fact become unemployed, Nilsson predicted, but "by 'unemployed' I do not mean unoccupied. Nor do I mean to imply that people will regard their unemployment as in any way undesirable. I merely mean that people's time will not be spent predominantly working for an income."

As to how people's time would be spent, Nilsson was not full of ideas, and the few notions he did have ranged from the dreadful ("volunteer or public-service activities") to the unrealistic ("artistic and creative pursuits") to the monotonous (the rise of "a 'Polynesian-type' culture").

Was that the final payoff of nanotechnology — mangoes and ennui?

To get a better grip on the matter, the NSG invited David Friedman, economist at the University of Chicago (and the son of Milton Friedman, the Nobel Prize winner), to give a talk on the subject at the January 1987 nano conference they were sponsoring.

But even Friedman was of two minds on the issue. The lesson from history, he said, was that technology did not mean massive unemployment.

"In this country, for example, since independence, roughly speaking, our population has gone up by almost two orders of magnitude. Throughout that time, aside from a couple of rather unusual years, the number of jobs and the number of workers never differed by more than ten percent. And I would think that just sort of with no theory at all, it seems very peculiar that you would have two independent numbers, each of which went up about a hundredfold during a period of two hundred years and that were virtually never more than ten percent apart."

What technology really did, said Friedman, was to increase productivity and create wealth. It gave people more free time.

"What happens," he said, "is that as technology improves, output per hour goes up. In a market economy people have the option of taking that improved output per hour in the form of more income — more real income — or in the form of more leisure. And of course over the last two hundred years people have done both on a large scale."

Nanotechnology would merely push this to extremes . . . and then some.

"No one will bother to charge for food," he said. "Food machines would be provided as a free amenity. They will be set up on street corners to commemorate dead spouses, just as water fountains are now. Raw materials, except for very large objects like planets, will cost almost zero."

A new car might cost something, but not much. "With Eric's technology we can make and market a new car for three seventy-five. That's three dollars and seventy-five cents."

This was pretty much the standard view, nanotechnology as the molecular cornucopia, and all was well and good.

Still, when it came time to say exactly what would fill a person's time in the nano age, when food machines were on street corners, when new cars went for $3.75, when only the larger planets were even remotely pricey, well, even Dave Friedman was slightly at a loss.

What would these people actually do?

"If it's important to human beings to feel that they're doing important, productive things, and if it turns out that in the world of a hundred years from now the important, productive thing you're doing is giving your wife a back rub, and you're both living off this phenomenal flow of income from your very small amount of property in a very, very productive world, you may feel useless.

"I can offer you no guarantees from economic theory against that."

The other item that Drexler had designed on his way to the warp drive was the nanotech computer.

His assemblers, after all, had to be controlled somehow, and they were not going to be managed by people, except indirectly. Their operations would be guided by programs, much in the way ribosomes were guided by DNA programs during the process of protein synthesis. This meant an army of nano-size computers.

The nanocomputer Drexler ended up designing was a rather startling machine — not because it was so advanced, but because it was, in a sense, so backward. It was not a miracle of electronics; it was not based on quantum effects or impalpable force-fields. Instead it was a molecular-level Tinkertoy computer, a mass of molecular rods and knobs, a machine in the truest and clunkiest sense of the term. His paradigm was not the integrated circuit but rather, as he said, "the Babbage machine."

The Babbage machine was the so-called Analytical Engine, a mechanical computer that had been designed and partially built by Charles Babbage in the mid-1800s. Much as Drexler's nanocomputer would be, the Analytical Engine was a masterpiece of mechanics, consisting of metal rods, rotating shafts, wheels, bearings, gears, cogs, ratchets, levers, stops, springs, and so on and so forth. It would have to be cranked and oiled; its inner workings would generate heat. Such a contraption was the very last thing anyone would expect Eric Drexler to take as a model when it came to designing a nanocomputer; nevertheless, he did.

There were good and sufficient reasons for it, though, the main one being simplicity: mechanical computers were far and away the easiest ones to analyze. Since his whole point was to offer a feasibility-proof, to demonstrate that computation was possible at the molecular scale, the most obviously workable design would be the best one to present, no matter how crude it was, and on this specific parameter the Babbage machine was ideal: "It was a computer that a child could understand," Marvin Minsky said.

A mechanical design had another advantage as well, namely that the physical behavior of the basic structures — the molecular rods, springs, and knobs — was easy enough to figure out, which meant that Drexler could come up with some actual numerical performance estimates. He'd be able to say what amount of data could be stored in what physical volumes, how fast such a machine would work, and how much heat it would generate.

The primary functional unit of any computer, large or small, was the logic gate, a device that provided a given output as a function of a given input according to a fixed rule. In electronic computers, inputs and outputs were binary states represented by distinct voltage levels — one voltage meant 1; another meant 0 — and the rule was contained in the circuitry involved. In Drexler's computer, binary states would be represented by the motion of a short, thin rod.

"A mechanical computer could transmit information by pushing and pulling sliding rods of molecular scale," he said. "Ones and zeroes could be represented by clamped and unclamped rods."

A rod in one position meant 1, in another position, 0. Movement of the rod was equivalent to the change of voltage in an electronic computer, and represented a signal. A molecular logic gate then consisted of a junction between two rods, each of which could block the movement of the other. When one rod was pulled back, the other one could move; when the first one was pushed forward, the other one was locked in place. Everything else was built up from that: logic circuits, memory cells, read-write heads, and so on.

The rods would be made of carbyne, linear chains of carbon atoms linked by alternating triple and single bonds.

"Carbyne is strong and stiff," said Drexler. "Since it is one atom wide, carbyne is at or near the limit of slimness."

The rod segments — each of which was exactly forty-four atoms long — slid back and forth within a housing that held the rod in place while permitting free movement along its length. Rods would be prevented from moving past their intended stopping points by "knobs," molecular bumps that were attached to the rods at regular locations.

Drexler's mechanical nanocomputer was a machine that could be built in the Big World with dowels and blocks. Implementing it at the molecular level, though, meant dealing once again with the idiosyncrasies of the atomic realm. In this case the graininess of matter presented no problem; in fact, it was even an advantage because it permitted the use of molecular groups, the "knobs," as stoppers. On the other hand, the thermal noise problem — *kT,* the source of Brownian motion — this now loomed as a major design constraint. Rods that constantly jiggled and vibrated due to heat, after all, did not seem to be choice candidates for secure information storage or precise signal transmission.

"The chief thermal noise problem is that a gate knob in a 0 state may move so as to allow a probe knob to pass, causing device-level errors," said Drexler. "A reasonable goal is to keep this probability below one in 10^{12}."

Which sounded difficult when considered in the abstract. On the other hand, if the thermal noise problem was really all that bad, then why didn't it interfere with the replication, transcription, or translation of DNA, which stored information at the molecular level just as his molecular computer would? So far as biology went, the

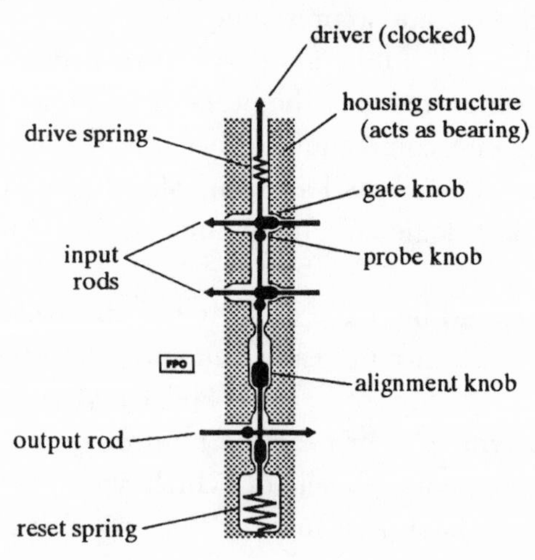

Molecular logic rods. (*K. Eric Drexler*)

molecular level seemed to be as safe and secure as it could possibly be, and if DNA molecules could manage the feat of information storage, copying, and transmission, then so could the molecules of his mechanical computers, especially if you helped them along with some clever engineering tricks.

Here there were at least two obvious tactics. Basically you could physically dampen out the thermal jigglings or else you could design the components so that they'd be insensitive to the magnitude of whatever jiggling there was in the system. And of course you could do both simultaneously.

A rod's transverse (crosswise) vibrations could be suppressed by making the rod housings large, stiff, and unyielding. Longitudinal (lengthwise) vibrations of the rod could be held within limits by putting knobs along the length of the rod to prevent it from traveling beyond a certain allowable distance.

You could deal with any remaining lengthwise vibrations by designing the system in such a way that the operating-displacements were larger in magnitude than the thermal vibrations were. In an electronic computer, for example, if a signal was equivalent to a change of one volt, then random variations of a hundredth of a volt would not affect operation, and would introduce few if any errors. The analogue of this for a mechanical computer was to regard as a signal only those rod movements that were bigger than the largest vibrations caused by heat. In that case, tiny thermal jiggling wouldn't affect computation because only bigger displacements of the rod would be regarded as a signal.

To Eric Drexler all of these schemes were entirely doable: all you'd need was enough atoms of the right elements in the right places — and nobody was going to run out of atoms. In his 1988 paper "Rod Logic and Thermal Noise in the Mechanical Nanocomputer," he calculated that thermally induced errors could be kept to the rate of below one in 10^{12}. Later, in his 1992 book *Nanosystems,* he did somewhat better than that, reducing errors to less than one in 10^{64}. "Error rates of this order," he said, "are negligible by almost any standard."

With nanocomputers of such description, you could extract some truly impressive performance from of the tiniest bits of matter.

"Detailed study shows that assemblers could build the equivalent of a large modern computer in about 1/1000 of the volume of a typical human cell," said Drexler. "Moving parts on this scale can be small and fast enough to make the computer faster than today's electronic machines. A desktop machine could then have more raw computational power than any computer in the world today. In fact, it could have more raw power than all the computers in the world today combined. In these terms — which imply nothing about intelligence — such a machine would have the raw power of a million human brains."

The reason why Dave Friedman hadn't been able to decide one way or the other in his anti-Luddite talk, conceivably, was that he hadn't realized the extent to which nanotechnology truly marked a break with the past. You wouldn't be "unemployed," he'd said, but all you'd actually be doing was giving your wife a back rub. Well, maybe his problem was precisely having taken prior experience as a guide. Doing that wouldn't help you out with nanotechnology because nanotechnology had no precedent, no parallel. There was really nothing remotely like it anywhere in past history. Never before were there machines that could manufacture anything that was physically possible, automatically, at nearly zero cost, thereby eradicating poverty, while at the same time curing all known illnesses, making people physically younger, enabling them to live for hundreds, perhaps even thousands of years, and, as an extra, added attraction, flying them off to the planets, to the stars . . .

No, there'd never been anything like that before.

And so when it came to imagining in concrete terms what life would actually be like after the Breakthrough, many in the nano clan tended to have the same problem. Other than for saying in general terms how wonderful it would all be when you didn't have to work, when you had all the time in the world to do everything you ever wanted . . . well, after saying that much, they really had no idea what to say next. And who could blame them, when past history was no guide?

But then came Jeff MacGillivray. MacGillivray had gotten a doctorate from MIT in physics, but he had an extremely broad mix of other interests, ranging across everything from far-out science fiction to the most arcane texts in the history of economics. He'd met Drexler back in the late 1970s at one of the early space-colony talks, and then attended all the usual parties, the space-habitat meetings, the nanotech lectures — the works. And of course he was a card-carrying member of the MIT Nanotechnology Study Group.

Two years after the 1987 symposium at which Dave Friedman spoke to mass confusion about the economic and social consequences of nanotechnology, the NSG held another conference and invited MacGillivray to rethink the whole subject. From the moment he stepped up to the podium it was clear that something new was in store, a rather different perspective on what what life would be like in the nano era.

Human values, for one thing, would undergo a profound change. Crass materialism — that poor flogged horse it had always been through the ages — would finally be dead and gone: people would place scant value on material objects in an age when automated assemblers gave them all the physical goods they could possibly want, and then some, for next to nothing. Material commodities would just be "around," like crabgrass.

Money, too, would no longer be worshiped the way it always had been; it would no longer be a badge of status. How could it be in an age when, one, everybody had tons of it, and, two, you didn't even need it, or very much of it, to acquire things? Essentially, the only material entities that would retain value in the nano age were land, which even nanotechnology could make no more of, and "special items of artistic merit," which was to say, artworks.

Since the work of the world would be done by machines — by the assemblers — there would be large amounts of unemployment. Nevertheless, there wouldn't be an unemployment "problem." Although you could certainly work if you wanted to, you wouldn't *have* to work for a living — that was the whole point. Unemployment would simply be a neutral and natural condition, the normal state of affairs, like the sky's being blue.

As for the four-billion-dollar question — What would people

be doing when they didn't have to work? — MacGillivray's answer was utterly simple: they'd be off enjoying themselves.

"We will have an entertainment society, not an information society," he said. "Self-directed people will pursue knowledge and entertainment for its own sake. Some will accumulate knowledge for the joy, satisfaction, and challenge of the pursuit. Others will take up artistic activities. Some will preserve and continue traditional and creative means of construction and production."

Art, performance, creation — these became the big themes of the future among the nano clan.

Nanotechnology, apparently, was greater nirvana. It meant instant and effortless satisfaction of every material want or need. It gave you everything, all on a platter. Matter had been overpowered; reality itself took on a new cast: it was controllable, plastic, malleable. It presented no further hindrance, no resistance to human will. The assemblers would provide.

Of course, there were a few problems with this. Such as how boring it might be to live in a world of that type, where nature itself, where the material world at large, offered few if any challenges. Was our paradigm to be the idle rich, people who'd whiled themselves away into drink, drugs, and depression?

MacGillivray himself postulated a certain pathology of the social order that would follow in the wake of the nanotech breakthrough. People would still want fame, status, and celebrity, but none of that would be obtainable from sheer wealth any longer because wealth had no significance in an age when everything was free, or nearly. Status would derive, instead, from success at ephemeral and counterfeit pursuits: one-upmanship in fads or fashions, for example.

"This will create a demand for copies of certain artistic designs, for tickets to performances of certain performers, and for close contact with fad leaders. Another group of people want to push other people around, and will probably be willing to pay for the privilege. This desire, merged with the desire to impress others, could lead to monumental displays of extravagance, employing enormous numbers of servants for short periods of time."

And as for the matter of insentient nature presenting no further hindrances, no resistance to human will, this didn't mean that such resistance would vanish altogether. Hindrance, resistance, and frustration would still be around, but would now be a product exclusively of *other people.*

Was this progress?

"The economics of production will change," MacGillivray said, "human nature won't."

That was the real problem: the perennial human clash and clatter would still be there afterward.

"There will be no end to the religious and ethical disputes which have plagued the human race throughout history: religious practices, abortion, and mind-altering drugs. The vast increase in standard of living will not make some people happy as long as any member of the human race has more income or wealth than them."

And so on and so forth throughout the litany of sins. People would still lust after each other's husbands and wives. Racial hatreds would persist unchecked. Smoldering political disputes would continue on unabated — or would even be magnified by some grand new nano methods, yet to be imagined, of rewarding your friends and getting back at your enemies.

And that wasn't even the worst of it. The worst of it was that while, beforehand, you had only one short lifetime to make a mess of things, after the revolution your opportunities for fouling things up would be virtually unlimited. You had a greatly extended human life span stretching out ahead of you, after all. You'd have thousands of years in which to commit faux pas, to have "misunderstandings," to offend people, and to watch your stupid mistakes pile up and be counted.

"All of these problems will be aggravated in duration," said MacGillivray, "and increased in frequency, by increased longevity — and in particular, by increased longevity in prime condition."

Then, too, the same boring people would still be around, like the rocks at Stonehenge, for untold centuries. *You'd* live longer, certainly, but so would the people you didn't like. Not just one failed marriage, but ten. Not just a few broken friendships, but hundreds. As your life went on — and on and on — the number of disappointments, botched relationships, and old scores to settle would rise uncontrollably, asymptotically, until . . .

Well, anyone could get the drift.

Bad as it was, there was yet a way out. "People in uncomfortable positions in such relationships may find it easier to move elsewhere and start over."

Nanotechnology would help here, too, of course. In the nano age, "elsewhere" meant "anywhere," courtesy of cheap and easy space travel.

"Space settlement will occur, spurred by such pressures as increasing population, the urge for adventure, and a desire to get away from past personal relationships."

And that, too, was part of the nanotech vision: people fleeing the aftermath of conditions they'd striven so hard to create.

Nothing new there, however: people built cities, formed ties, entered marriages, and then turned around and escaped them in droves. That was just human nature.

Nobody ever said (least of all Drexler) that nano would solve every problem. It could only solve physical problems, the *easy* problems, like poverty, aging, and disease. It couldn't turn people into saints or angels — but was it fair to hold that against it?

Drexler's papers on "rod logic" struck a responsive chord in A. K. Dewdney, the "Computer Recreations" columnist for *Scientific American*. His January 1988 piece was devoted entirely to Eric's molecular computers.

"I've got a soft spot in my heart for new and interesting machines," Dewdney confessed. "Drexler's machines offered a peek into the future."

Dewdney laid out Drexler's whole program, everything from molecular gears and bearings ("a key device in any conceivable nanotechnology is the bearing," said Dewdney) to assemblers ("small machines that would guide chemical bonding operations by manipulating reactive molecules") to practical applications in health and medicine ("wherein molecular computers control tiny circulatory submarines").

"Microchips are small," said Dewdney. "Their components are scaled in the micron (millionth of a meter) range. Drexler now asks us to consider a computer that would fit inside a single silicon transistor!"

So Dewdney considered it. One of the main advantages of such tininess, he discovered, was speed. "They are not at all slow," Dewdney said of Drexler's carbyne rods. "At the atomic scale such rods take only about 50 picoseconds (trillionths of a second) to slide."

Dewdney examined Drexler's "rod logic" designs in detail. He traced out the rod-logic workings of a sample numerical calculation, showing exactly how it would proceed down in the molecules. In theory, at least, he found no problem with any of it. Drexler seemed to have thought of everything. His system ought to work as advertised — provided, of course, that it could actually be built.

"Those carbyne rods have to be doable," he said, looking back on his column a few years later. "They have to be buildable. But we don't know anything of what is or is not buildable in the nano realm. I remember thinking to myself, 'Jesus, what kind of complicated fabrication machines they're gonna have to have to build these little computers! They're gonna be mildly fantastic. Thank God I don't have to worry about *that!*'

"We won't really know how some of these things are going to work until we get them going," he added. "Molecules are very tricky. For example, calculating friction, you've got to build a computer simulation model and see how the repulsive electrical forces actually work out. Drexler's done some of this already; the question is how much more he has to do before we can move toward building actual bits and pieces."

Faraway as it was, all of it still sounded entirely possible to Kee Dewdney.

"Although nanotechnology is currently hardly more than a gleam in the eye of Drexler and a handful of other scientists, it has a curious ring of inevitability. The future it implies seems an order of magnitude more wonderful than anything in science fiction."

So Drexler now had, in addition to his molecular gears and bearings, a little molecular nanocomputer.

This stuff could really be done. You could actually build an assembler provided that you had enough of those gears, bearings, and nanocomputers — and provided, of course, that you had an assembler . . . or some other way of putting those parts together.

11

"Three Little Gears"

It wasn't until he saw the *Omni* cover story about nanotechnology that Barry Silverstein decided that he really had to read *Engines of Creation.* A copy of the book had been sitting on his "to read" pile for a couple of months, and when he finally got to it he thought it was the most optimistic assessment of the future of humanity he'd ever seen.

Drexler's proposed technology, he concluded, would be nothing short of miraculous. It promised you near-immortality. It promised to solve completely the problem of economics: "buying things" would no longer be an issue once there was plenty of everything available for almost nothing. It promised health, youth, space travel.

Silverstein didn't have much in the way of a science background but he bought about a dozen copies of *Engines* to give away to science-minded friends of his — doctors and such — so that he could talk to them about it. But few of them really cared for the book: too far out, they said; too hard to believe.

But Silverstein was so much impressed with it all that he resolved he had to meet this Eric Drexler character, and when he eventually got in touch with him through the Foresight Institute he learned that Drexler would back in the Boston area soon to give another set of lectures at MIT. A few weeks later, on a snowy evening in January, they met for dinner at Locke-Ober, the politician's hangout in downtown Boston.

Silverstein was quite struck by Eric Drexler: he seemed so reasonable, so competent, so low-key. No flakiness there at all. He asked Eric how things were going, generally, and Eric said things would be all right if only he hadn't had to keep his head above water by giving lectures to the Ladies' Girdle Manufacturers Association, or whatever, quaint groups who invited him in mainly for the shock value — "Come hear this crazy scientist," and so on. He'd rather be doing his usual work, theorizing, writing, giving technical talks.

Silverstein, who'd made a small fortune in cable television and other ventures, and who lived in upstate New York not far from the town of North Pole, was in the habit of giving money away to worthy causes of various types, and he thought that Eric Drexler was by far the best and worthiest candidate in memory. Shortly thereafter he was sending him money on a regular basis, monthly checks to cover living expenses, research, and to help out with the Foresight Institute. Later, he'd encourage Eric to go back for his Ph.D., and he'd finance that, too. In all, Barry Silverstein would end up sending Drexler a total cash amount that reached well into the six figures.

Not that any of it was reflected in the lifestyle of Chris and Eric. They were, after all, two of the most future-oriented persons in history and weren't the types to blow money away in the mere present. They did, however, move out of their Redwood City apartment and into a house, which Chris had been wanting to do for some time. A modest two-bedroom ranch house on a Los Altos cul-de-sac, it would remain furnished in early graduate-school for years to come.

After a while they also broke down and bought a new car — a used car, actually, a Toyota Camry station wagon — but that was about it as far as the big spending went. Neither of them were much for taking vacations, either, and once when Eric won an all-expense-paid trip for two to Hawaii they hemmed and hawed for a full two years, until the prize offer was just about to expire, at which time Eric's father persuaded him that, "Look, Eric, a little

R&R really might not be a bad idea just about now" — and so they threw up their hands and went.

Mostly both of them just worked like crazy, Chris for Foresight and Eric on assorted new nano projects, the most important of which was the Stanford course he was scheduled to teach. The chairman of the computer science department, Nils Nilsson — an initial skeptic about nanotechnology, but quickly converted — asked Eric if he'd offer a formal course on the subject, and after some initial hesitation he agreed. Teaching it, after all, would force him to get his thoughts in order, to systematize the material. It might even get him moving toward writing his technical text.

So during the spring semester of 1988 Eric Drexler taught Stanford course CS404: "Nanotechnology and Exploratory Engineering," the world's first and only college course in the subject. It drew an audience of about fifty students, one of whom, because there were so many others blocking the doorway, climbed in the window. Eric took the class through all the old stuff: molecular gears and bearings; thermal noise and the thermodynamics of computation; protein design as a possible pathway to first assembler; the dangers and benefits; the likely social implications. But there was some new material in there, too, including an engineering design for a general-purpose mechanical assembler, a contraption that looked like nothing so much as a nanoscale industrial robot arm, which was in fact how Drexler often described it.

Also making its first formal appearance was the so-called single-point failure assumption — the premise that molecular components would be regarded as having failed if they underwent a single unplanned chemical transformation anywhere in their structures. If even so much as one lone atom was out of place, then the device in question was considered to be nonoperational — which was about as extreme as you could get in the annals of engineering-design conservatism.

Like many others who studied the subject, the main difficulty the students had with nanotechnology was that there seemed to be no known way of making it happen.

"The key problem here is that nobody knows how to build systems to atomic specifications," said Martin Rinard, who took the course. "Drexler reduced this to the problem of constructing a general-purpose assembler consisting of assembler arms mounted on the base of a cylindrical housing with a nanocomputer for control. To build this assembler, Drexler envisions a series of progressively more complex assemblers, each of which builds its successor. In such a scenario, the first assembler might be made of protein molecules using conventional protein-synthesis techniques.

"Although this bootstrapping sequence seems intuitively plausible," he added, "it is so sketchy that it is hard to say whether or not it can be done any time in the foreseeable future."

This sort of point would be made again and again — that

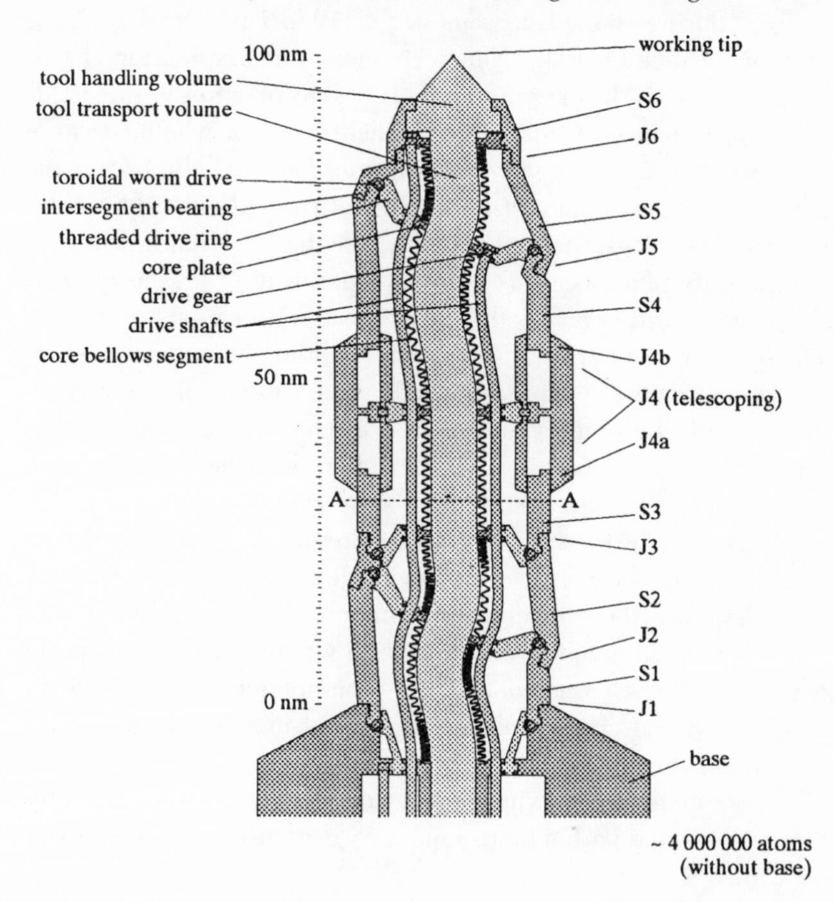

Molecular manipulator arm, or "assembler." (*K. Eric Drexler*)

Drexler's proposed devices might work if they could actually be built, but who knew if they really could or ever would be, since, as yet, there were no tools in existence with which to construct them. That had been the attitude of Sheldon Glashow and George Whitesides, for example, toward Eric's molecular gears and bearings: How do you *make* them?

Drexler, though, had already answered that question in his *PNAS* piece, half of which, roughly, was given over to just that subject. You'd make these gears, bearings, and other machine parts by protein engineering. You'd design new proteins that would fold up into components that would then assemble themselves into a larger, workable molecular machine.

Protein engineering had not come about as yet, true enough, but that was no barrier to Drexler's overall argument. The whole point and purpose of his work, as he'd told the class during the first meeting of the Stanford course, was "exploratory engineering," by which he meant theorizing about designs and devices that were, by definition, beyond what was then possible, "beyond the reach of implementation using present tools and resources." It meant regarding natural law, instead of current factory or lab techniques, as the guide to what was potentially achievable.

If his planned devices were in fact worth making, if they indeed offered the range of possibilities that he held out in promise, then their benefits would motivate people to invent the tools that would make the devices. Fundamentally this was a means-end question: if the end was worth attaining, then you devised the means for reaching it. You had to go out and create the future, not just wait around and let it happen.

What a surprise, then, when a brand-new nanotool, totally unrelated to protein engineering, just happened to happen along. And how fitting that the device in question — the scanning tunneling microscope, or STM — resembled, at least in gross outlines, the sort of master-slave device that Richard Feynman had described in 1959.

The STM, in fact, was a far simpler tool than even Feynman had dreamed was possible. With it you wouldn't have to go through the staged sequences of ever-smaller machines that he'd laid out in

"Room at the Bottom." Instead you'd move atoms directly, immediately, without all the intermediate downsizing. You'd just sit at the lab bench, give some instructions to this surprisingly simple machine in front of you, and lo and behold a tiny mechanical arm, quite like the tiniest slave hands in Feynman's original scenario, would move just slightly, taking one or more individual atoms with it.

Cal Quate first heard about the scanning tunneling microscope in April 1982 as he was winging his way across the Atlantic. A professor of applied physics and electrical engineering at Stanford, Quate was on his way to a scientific conference in London when he started reading the issue of *Physics Today* that he'd brought along with him. "I think we were over Iceland when I opened it and found a report on a new form of scanning microscopy being developed in Zurich," he recalled.

The report, "Microscopy by Vacuum Tunneling," told of a new type of microscope that was just then being developed at the IBM Zurich Research Laboratory, by the two physicists Heinrich Rohrer and Gerd Binnig. The microscope worked by a novel principle: no illumination of any type was required for it to "see" things; no light, no X rays or anything else was to be shined on the specimen. Instead, this new device worked mechanically, by touch, by actually feeling its way around an object and sending back messages from which a visual picture of its size and shape could be constructed. It was like getting an image of a river bottom by taking manual depth-soundings — Mark Twain style, throwing a plumb bob over the side — and then plotting out the various measurements by height.

The second unusual thing about the STM was its truly exquisite physical sensitivity. According to the *Physics Today* report, the device was capable of feeling its way along a surface with "a precision of one or two tenths of an angstrom." That was almost surreal. A typical atom was about an angstrom or so across and here was a new instrument — one that worked mechanically, by touch — taking measurements one tenth that size.

Which could only mean that with this device you ought to be able to "see" the atoms themselves, as indeed happened soon enough. "Wavy structures of variable periodicity" were showing up in "flat" surfaces, the *Physics Today* report said.

Those "wavy structures," Quate thought, could only be the hills and valleys of atoms.

That was it for Cal Quate. "In London, I changed my travel plans and went to Zurich."

It was entirely logical, in retrospect, that the scanning tunneling microscope would be invented by the world's biggest computer company, IBM. Computer chips, after all, were tiny structures to begin with and they only kept on getting smaller and smaller, ever more tightly packed with miniature electronic devices. To keep track of what went on down there among the various circuits and switches you had to come up with ever-better ways of seeing the things.

In the spring of 1978 Gerd Binnig, who was then a grad student at the University of Frankfurt, was having a discussion with Heinrich Rohrer of IBM Zurich about how to get accurate pictures of computer-chip structures. Binnig suggested making use of the phenomenon known as vacuum tunneling, which was a concept from the realm of quantum mechanics. Under normal conditions the surface electrons of a given piece of matter stayed where they were, which was to say, embedded within the surface itself. Nevertheless, if you placed two pieces of matter close together, a few stray electrons would spontaneously cross the gap between them. This was known as "tunneling" because the gap constituted a theoretical "barrier" that the electrons would have to dig their way through.

Binnig suspected that you could make a microscope based on the tunneling principle. If you took an extremely sharp needle tip and passed it over a surface, then you could monitor the rate at which electrons tunneled between the surface and the tip. The incoming electrons would produce a measurable electrical current in the sensing apparatus: the narrower the gap, the stronger the current, and vice versa. This meant that if the needle tip was held at a constant height as it scanned back and forth across the specimen, then the rise and fall of the electrical current would represent the physical ups and downs of the surface.

Binnig came to IBM Zurich in November of 1978 and immediately started working with Rohrer. Three years later, the two of them had produced a working device, one that turned out to have a far better resolution than either of them would have forecast. The vacuum-tunneling phenomenon was extremely sensitive to the size of the gap: a tiny change in the opening resulted in a big jump in the flow of electrons across it.

"A change in the distance by an amount equal to the diameter of a single atom," they noted in one report, "causes the tunneling current to change by a factor of as much as 1,000."

After working with the device for only twelve hours, Binnig and Rohrer were getting images of features that were even *smaller* than atoms. "Our microscope enables one to 'see' surfaces atom by atom. It can even resolve features that are only about a hundredth the size of an atom."

Soon they noticed something even more amazing, that the needle tip was accidentally picking up stray atoms and moving them from place to place on the surface. This was due to the imperfect control they had over their machine: sometimes the needle tip would actually nose-dive right into the sample and bring an atom or two away with it.

Well, what was this but *moving atoms?*

That had never been done before, certainly not mechanically, and certainly not just an atom or two at a time. Erwin Müller, it was true, had fleetingly isolated single atoms back in 1967 with his so-called atom-probe field-ion microscope, a variation of the field-emission microscope with which he first saw the atoms. This new atom-probe device of his worked as the earlier one did, by emitting bunches of atoms from a hot metal cathode at the base of a TV tube and letting them fly up toward the viewing screen, upon which they left their traces. But between the cathode and the screen he now placed a barrier in which there was an extremely tiny hole.

By sighting through the hole, Müller was able to single out a specific atom — "one atom, selected at the discretion of the operator" — fire it off the needle tip (together with many others), and allow it to pass through the aperture. For the short space of time it took the atom to pass through the hole and whiz through the sensing apparatus — a cosmic eyeblink — the atom was, in a sense, "isolated."

Impressive a feat as it was, the process was the atomic equivalent of shooting buckshot at a keyhole and getting only one piece of shot to pass through. Binning and Rohrer's atoms, by contrast, were not careening along in flight but, until moved by the STM tip, were stationary and in place on the surface. And their atoms, unlike Müller's, were not deployed in a large-scale field-emission process but rather were manipulated individually and mechanically, by push, just as you'd move a chess piece from one square to another, by touching it and giving it a shove.

Thus far Binning and Rohrer had moved single atoms only by accident, but the possibility remained that you could learn to do this intentionally, in which case their STM would be an atomic-manipulation tool, exactly what Feynman had imagined in 1959 when he wondered "whether, ultimately — in the great future — we can arrange atoms the way we want; the very *atoms*, all the way down!"

Feynman's "great future" arrived twenty-seven years later, in 1986, when three researchers, Russell Becker, Jene Golovchenko, and B. S. Swartzentruber, of AT&T Bell Laboratories in Murray Hill, New Jersey, announced that they'd made "an atomic-scale modification of the surface of a nearly perfect germanium crystal."

They'd created a needle tip so sharp that it was "terminated by a few or perhaps even just a single atom." The tip was controllable in all three dimensions to an accuracy of .1Å — a tenth of an angstrom, a fraction of an atomic diameter. They'd taken the needle tip, placed it above a region whose atomic structure they'd already mapped out, and applied a small electrical voltage. When they looked again, a tiny bump had appeared out of nowhere, a "small isolated new protrusion in a previously pristine region of the surface." The bump was a single atom, quite probably the very one that had constituted the point of the needle.

In an article entitled "Atomic-Scale Surface Modifications Using a Tunnelling Microscope," published in *Nature,* the authors now claimed a world record for themselves: "We believe this to be the smallest spatially controlled, purposeful transformation yet impressed upon matter." They also claimed to have plumbed a final and irreducible limit of nature: "We argue that the limit set by the discreteness of atomic structure has now essentially been reached." And last but not least, they claimed to have demonstrated atom-by-atom control of matter, precisely what Drexler had been predicting since 1977: "We hope to have shown that man can now manipulate a few chosen atoms for his own purposes."

The authors made no claims, however, as to what those "purposes" were.

The AT&T world record was shattered in less than a year by three researchers at the IBM Almaden Research Center, in San Jose. John Foster, Jane Frommer, and Patrick Arnett, working with an STM of their own, had noticed a little "bump" appearing every so often here and there on the test surface. A bump at that level of magnification, they reasoned, could only be "a single molecule or fraction of a molecule."

They could deposit a single molecule on the surface, and they could also remove it again: "The reverse manipulation, the removal of pinned molecules, has also been demonstrated."

And in fact they went even one better than that, actually breaking a molecule in half, separating it into its two atomic components: "We can remove a portion of the pinned molecule, effectively performing transformations on single molecules using the tunnelling microscope."

That was something new in world history, the appearance of a mechanical chemistry, one in which molecular alterations were made not on vast masses of molecules in solution but on single molecules, individually, which were ripped apart by physical force. It was an example of Drexler's proposed "mechanochemistry."

Having beaten the AT&T team at their own game, the IBM group now alleged the world's record was in fact theirs since *they* had produced "what we believe to be the smallest spatially localized change of matter achieved in a laboratory." Their mark, after all, was "about half the size of the mark made" by the AT&T group!

Which meant that yet another of Feynman's hopes had been realized. "Let's have a competition between laboratories," he'd said in 1959. "Let one laboratory make a tiny motor which it sends to another lab which it sends back with a thing that fits inside the shaft of the first motor."

Thirty years later, the labs had gone far beyond tiny motors. They were now competing at the level of atoms.

Still, the question remained what the point of it all was. The AT&T bunch had spoken of man's being able to "manipulate a few chosen atoms for his own purposes." The IBM troupe now repeated the sentiment, saying that "tunnelling microscopy is on the threshold of a revolution in manipulating atoms and molecules for a variety of purposes." But they, too, never said what any of those "purposes" were.

All of which meant that a rather marvelous new situation was now occurring in the greater world of science and technology. Feynman's predictions of a far-off era of atomic manipulation had suddenly come true, and much sooner than he'd ever imagined. The instrument that had worked the miracle, furthermore, was arguably similar to, if even more science-fictional than, the master-slave device proposed by Feynman, which had in turn originated with the science-fiction writer Robert Heinlein.

Feynman, on the other hand, had never come up with any practical "uses" for the tiny machines he so much loved, and neither had the researchers who were now in the process of developing the instruments and techniques with which those machines could be made.

But when Drexler not only showed how the machines could be constructed, but also spelled out in graphic detail the whole range of uses and purposes, good and bad, to which they could be put, critics responded by doubting that any of it could ever be done and charged him with purveying "science fiction."

If better evidence were needed that a paradigm shift was now in the works, it was hard to imagine what it could be.

* * *

Although Drexler had talked about designing proteins and building them to order in his 1981 *PNAS* piece, neither he nor anyone else had any firm notions as to how far off in the future such a development was. Given the complexity of the protein-folding problem, however, it could conceivably be a long time away. How agreeable it was, therefore, when during the winter of 1987/88 the world's first artificial protein was designed and actually built in the lab.

The event took place at E. I. du Pont de Nemours and Company, in Wilmington, Delaware, where a small group of du Pont researchers headed up by William DeGrado proposed to design and build a modest protein: it would consist of four distinct corkscrew-shaped sections (helixes) connected together by a series of loops. In an outline of their plans published in 1987, DeGrado cited "Drexler 1981; Pabo 1983" and said that the research team was taking Drexler's "inverted approach" as their starting point: "If it is not possible to predict the three-dimensional structure of a protein from its sequence might it not be possible to do just the inverse? Could one begin with a reasonable three-dimensional structure and then design a peptide sequence that would fold into this structure?"

Drexler's "backwards" approach was natural enough, since it approximated ordinary Big World design and construction methods. When architects designed a house, for example, they never started out by considering the finer points of nails, screws, and wood glue and building up from there. Just the reverse: they began with an overall conception of what the house should look like and only then specified the component parts that would go into it.

DeGrado and company worked in similar fashion. Taking naturally occurring proteins as models, they started out with a picture of what the final, folded protein would look like, and only then figured out what were the likely subcomponents that would, they hoped, produce it. The key to it all was the known fact that certain amino acids had special affinities for producing proteins of distinct shapes.

"It has long been known that certain amino acids have clear-cut preferences for adopting a given secondary structure," said DeGrado. "Thus it is now routinely possible to design peptides that incorporate single secondary structural units. A logical next step would be to design secondary structures that can pack together to form a globular protein with a predetermined three-dimensional structure."

Protein molecules were so morphologically complicated that they had at least four separate levels of organization: primary, secondary, and so on. The primary level consisted of the particular sequence of amino acids that made up the protein. The secondary level was the way in which that chain of amino acids arranged itself in space geometrically, and here there were four different possible arrangements: alpha helixes (corkscrews), beta sheets (flat, wavy ribbons), turns, and loops. The tertiary structure was the next-higher-up level of organization, and referred to the manner in which the various helixes, loops, and whatnot combined to generate the final, three-dimensional folded shape.

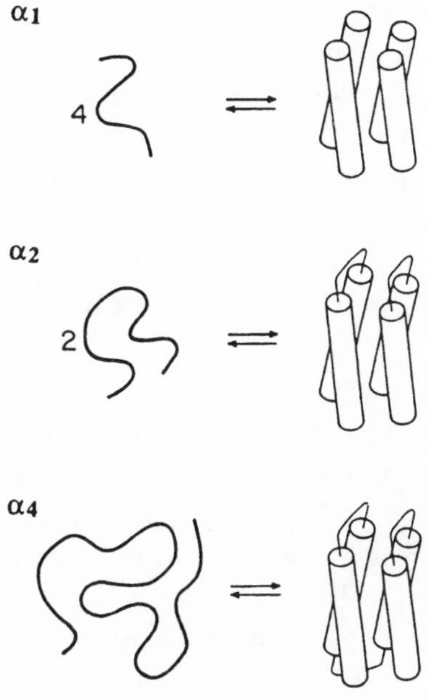

Design sequence of the world's first artificial protein. (*William DeGrado*)

DeGrado and his group decided to build a protein molecule that consisted of four corkscrews (alpha helixes) each one connected to the next by a loop — a molecule which, when made, would look like four sticks of dynamite wired together in series. Supposedly, the correct amino acid sequence would produce the four alpha helixes and the three loops, all of which would assemble themselves into the proposed dynamite-stick configuration.

That was the design side of the picture; actual synthesis was another matter. Physically, the protein would be constructed by *E. coli* bacteria, whose ribosomes would be fed the precise string of DNA nucleotides that encoded it. The bacteria's ribosomes would read the DNA nucleotides and string together the amino acids in the prescribed order, bringing the intended protein into existence.

Bill DeGrado, Lynne Regan, and their du Pont colleagues established what they thought was the right DNA sequence and then produced it in an automated DNA synthesizer — the so-called gene machine.

Gene machines had only recently been invented: they were devices that manufactured short segments of DNA — they actually made DNA — out of a supply of bottled chemicals. What's more, they manufactured it to order. It was simply a matter of knowing the base-pair sequences you wanted, supplying the right chemicals, and pressing a few buttons. Soon, out came your predesigned DNA segments, correct to the last base-pair.

Both the idea and the implementation of a gene machine bore more than a passing resemblance to Drexler's "meat machine," a device that would synthesize meat from a stockpile of chemicals. But while the meat machine was still imaginary, the gene machine was real enough, and by 1983 a firm called Applied Biosystems, of Foster City, California, was selling desktop DNA units for $42,500 apiece.

Essentially it was a keyboard, display screen, and rank upon rank of bottled chemicals, all of them connected up by the appropriate wires, tubing, and microprocessor chips. A trained technician could operate the machine after a few hours of instruction. He or she followed the prompts on the screen, keyed in the desired DNA base-pair sequence, and stored it all in computer memory.

The machine did the rest. The manufacturer estimated that a twenty-nucleotide piece of DNA could be synthesized for about fifty dollars per nucleotide.

The du Pont group created their DNA sequences in a gene machine and then introduced them into an *E. coli* culture. Everything went according to plan and by the spring of 1988 the team had created their own little artificial protein, which folded up exactly as it had been designed to, into "four identical, designed helices connected by three identical, designed loops."

Their report, published in *Science*, said: "The complete α_4 protein is the first example of a designed protein that has been shown to adopt a folded, globular conformation in aqueous solution."

Their synthetic α_4 protein was actually "better" in some ways than the naturally occurring proteins they'd taken as models: "The folded conformation of the protein is extraordinarily stable. Also, the interior of the protein is more regular and perfectly hydrophobic than in natural proteins."

To Eric Drexler, this was progress. "It's a major landmark, which I'm very excited by," he said shortly afterward. "And it marvelously confirms the predictions I made in 1981."

DeGrado, for his part, soon began speaking like a dyed-in-the-wool nanotechnologist. Protein engineering, he said, "will allow us to think about designing molecular devices in the next five to ten years. It should be possible ultimately to specify a particular design and build it. Then you'll have, say, proteinlike molecules that self-assemble into complex molecular objects, which can serve as machinery."

In 1986, the year *Engines of Creation* was published, no actual mechanisms existed for building the fabled "assemblers." Now, suddenly, a mere two years later, two such mechanisms had been invented and had been found to work: one was the "inverse" protein-engineering technique that Drexler had himself proposed; the other was the scanning tunneling microscope.

So much for the question "How do you *make* them?"

* * *

It was inevitable that nanotechnology would arrive on the electronic bulletin boards of various computer networks. First there was *Omni* Forum, an interactive conference run on CompuServe following the magazine's "Nanotechnology" cover story in November 1986. The NSG regulars Dave Lindbergh, Kevin Nelson, and Dave Forrest were on-line to answer what by this point were getting to be entirely predictable, run-of-the-mill questions. There was, for example, the inevitable quantum-uncertainty objection:

> One thing that this field will have to contend with is quantum mechanics. . . . It will be difficult to design machines to perform to specs in a realm where probability rules over certainty.

The NSG answer to which was:

> Quantum mechanical uncertainties do not prevent the extremely sophisticated machinery of our own cells from working. . . . We're not talking 'warp drives' here. This is for real.

Quantum uncertainty, it turned out, had a far worse reputation than it actually deserved. It was another one of those things everybody "knew" about, but which few understood in any detail; in fact it was the rare person who could tell you what the physical magnitude was of a given particle's positional uncertainty, or could even say whether that uncertainty was to be measured in millimeters, nanometers, or miles. Nevertheless, everyone "knew" that quantum uncertainty, whatever its magnitude, was bad enough to derail Drexler's whole program.

Physicists, on the other hand, were aware that quantum uncertainty posed no barrier to the existence of molecules. "The uncertainty principle would not forbid a reasonably precise location of an atom because an atom is sufficiently heavy," said Robert Walker, a Caltech physicist. "And as a matter of fact all of the theories of molecular structure and so on depend on the relative locations of atoms. Molecules wouldn't be possible at all if precise locations of atoms were forbidden by the uncertainty principle."

But if molecules were allowed by the uncertainty principle, then why wouldn't molecular machines, which, after all, would be much bigger than individual molecules, also be allowed?

In Eric Drexler's eyes, although quantum uncertainty was real enough, the magnitude of the effect as it pertained to atoms (as opposed to electrons and other subatomic particles) was so small — measured in the tenths of an angstrom — that for most practical purposes it was negligible. His viewpoint had been corroborated in spectacular fashion by Hans Dehmelt's pet particles, Astrid and Priscilla, whose locations inside particle traps were certain in the extreme, and for months at a time.

Drexler's other main point about quantum uncertainty was that to build nanomachines you didn't in fact *need* infinitely precise placement of atoms. In any sort of engineering you only needed an accuracy sufficient to the purpose at hand, and for the purpose of building nanomachines the uncertainty in the position of an atom was so very tiny — a small fraction of an atomic diameter — as to be acceptable in almost any context.

Anyway, the uncertainty issue came up again and again on the computer networks, but so did lots of other related topics. After a while there was so much back-and-forth nanotalk on the bulletin boards, discussion groups, and so on, that one nanofan formed a breakaway group of his own and reserved it exclusively for discussion of nanotech stuff. This was John Storrs Hall, who on the networks signed himself off as JoSH.

JoSH had read *Engines of Creation* shortly after it came out.

"I was highly enthused," he recalled. "I spent an awful lot of time at that point reading up on chemistry and microbiology, which had been gaps in my education."

A Rutgers grad student in computer science and artificial intelligence, JoSH concluded early on that he wanted to meet Eric Drexler — "From reading the book I had the notion that he was a kindred soul, the sort of person that I could just talk to and understand" — an expectation that was confirmed when he finally met up with him at an AI conference.

"It was sort of like meeting your long-lost older brother," JoSH recalled afterward. "That made my year, I guess."

JoSH was the moderator of a politics discussion group on

USENET, and at one point he decided that the volume of talk about nanotechnology was great enough that the subject merited its own group. To start a new one, though, you had to put it to a vote among the users.

"The rule was that you had to get a hundred more yes votes than no votes, and then the group could be created. It turned out that the nanotechnology group got one of the highest ratios of yes-to-no votes of any group ever. The tally was something like three hundred to seven."

That was in the spring of 1988. The newsgroup was called "sci.nanotech," and JoSH himself posted the first message:

> Welcome to the nanotechnology newsgroup. This message is largely a test of the slightly arcane software we have set up to handle the postings.
>
> sci.nanotech is for the discussion of molecular technology. As Herbert Hoover might have said, "De novo enzyme design is just around the corner." At first, I'll take Engines of Creation as a definition-by-example of the subject matter considered appropriate for the group. This will expand or contract dynamically depending on volume and quality of the postings.

Less than a year into the operation, after hundreds of pages of discussion had been posted about every conceivable nanotech topic, it seemed that every skeptical objection had been instantly met by the fans.

> The thing that bothers me most about nanotechnology is the power source; how do you supply power to these molecule-sized machines? What do they use for fuel? Even on a molecular level, you still have to obey the basic laws of physics.

The answer to which was:

> Did you know that there are trillions and trillions of invidious little nanomachines already running rampant in our biosphere? I am speaking, of course, of bacteria and viruses. These little critters require energy to operate. Last time I checked, no one has had to build power plants for the purpose of keeping the dear little things revved up. So maybe if we're REALLY smart, and think REALLY hard, and eat our Wheaties, we just MIGHT find a way to power our own nanomachines.

Soon it was clear that nanotechnology would be talked about more on the nets than in the scholarly journals. At one point Drexler claimed that if he ever made a big mistake anywhere in his papers or books, it would be all over the nets in a matter of minutes.

Every so often a newcomer to the sci.nanotech newsgroup would bring up the "three little gears" business.

> I keep seeing the pictures of three little gears flashed by nanotechnology proponents.... The three gear picture must be regarded as a stunt.

How true that was. These three little gears, indeed, were getting to be extremely famous. They were state-of-the-art prototypes in a competing craze known as microtechnology.

Microtechnology was concerned with making mechanical components that were "small" but still essentially macroscale objects. The three little gears, for example, were each roughly the diameter of a strand of human hair, some fifty microns across, whereas the devices that Drexler contemplated were, on average, more like fifty nanometers across.

In fact, compared to the devices that Drexler was envisioning, the three little gears were not even in the same mental, moral, or physical universe. They were Big World creatures one and all, whopping monsters composed of millions of molecules, and many trillions of atoms.

"The micron scale," Drexler said, "is volumetrically 10^9 times larger than the nanometer scale. Confusing microtechnology with molecular technology is like confusing an elephant with a ladybug."

The distinction was lost on many outsiders, who tended to package the two technologies together because both of them, after all, dealt with things that were "small." But in truth the micro and nano realms, and the technologies rooted in the respective size scales, had nothing of any importance in common. For one thing, microtechnology was still a bulk manipulation of matter.

"Microtechnology dumps atoms on surfaces and digs them away again in bulk, with no regard for which atom goes where," said Drexler. "Its methods are inherently crude. Molecular technology, in contrast, positions each atom with care."

Another difference was that whereas microtechnology started out with big blobs of matter and then whittled them down to size, nanotechnology would go in the other direction, starting out with individual atoms and then building up.

"In microtechnology, the challenge is to build smaller," Drexler said. "In nanotechnology, the challenge is to build bigger."

But even that wasn't the main difference between the two. The distinguishing feature of nanotechnology was that it gave you control over chemical reactions through the forcible, mechanical placement of individual molecular groups. That was nanotechnology's whole raison d'être; it was the feature responsible for the prospect of molecular manufacturing, for the promised "complete control of the structure of matter." Microtechnology, whatever its virtues, did not offer precise molecular positioning as one of its benefits, which meant that it could no more give you complete control over the structure of matter than it could give you invisibility.

And in fact it couldn't give you much of anything else, either. This was the view of its pioneers, anyway, or at least some of them, who quite cheerfully admitted they had no idea of what their tiny trinkets would be good for, if anything: "To some extent we have an answer, and we're looking for a problem," said Jeffrey Lang, a microtechnology developer at MIT. The point was confirmed by a *New York Times* microengineering story that ran under the headline "New Generation of Tiny Motors Challenges Science to Find Uses."

That was the challenge, all right: finding uses and purposes for the goddamn things. No one but Eric Drexler, apparently, had any use for "uses."

As for the three little gears themselves, the tiny, toothed marvels had been produced up at AT&T. They'd been made essentially the same way computer chips were, by depositing layers of silicon and then selectively etching matter away in predefined patterns, leaving you at the end (in the case of the gears) with a freely moving structure.

Magazines and newspapers the world over just absolutely loved those three little gears. They ran pictures of them, ran endless stories about them, in articles that overflowed with bad jokes about blowing away whole rafts of the things just by sneezing, breathing, or maybe even thinking too hard: "One sneeze could ruin your whole production line," and so forth. ("Sneezing isn't the problem," said one iconoclast. "Inhaling is the problem.")

Glamorized as it was by the mags and papers, microtechnology was still but one step down from Bill McLellan's 1/64-inch electric motor. He'd never found a "use" for that, either. "It could be employed to run the merry-go-round for a flea circus," he joked.

AT&T's "Three Little Gears." (*AT&T*)

So you really had to wonder about those three little gears. They were photogenic. They were cute. They were charming.

They just didn't serve any known purpose.

You could imagine them, those three little darlings — perfect angels! — you could imagine them in a Disney movie. That would be a use: send them to Hollywood. You could imagine them in a Busby Berkeley number, dancing and singing. Or Gilbert and Sullivan, even . . .

> Three little gears for tools are we,
> Pert as a little gear well can be,
> Filled to the brim with gearish glee,
> Three little gears are we.

12

Captain Future

Finally and at long last, Drexler and his creation were getting some institutional recognition. In the fall of 1988, the same year he taught the Stanford course, a group of graduate students at the University of Texas at Austin began a year-long study of the scientific prospects and the political, economic, and social consequences of nanotechnology. The project, headed up by Susan Hadden, a professor at the Lyndon B. Johnson School of Public Affairs, came about after Roger Duncan, an old friend of Hadden's from Austin city politics, read *Engines of Creation*. Duncan, a former Austin city council member, decided that if Drexler's scenario really worked, human life would never be the same again.

Duncan talked with some of his friends and found out that their response to nanotechnology fell into two categories. There were the technoids who felt that, Oh yes, this was going to happen, all right, but whose faces turned blank when the issue of social impact was raised: What social impact? And there were the politicians who felt that, Oh yes, there'd be a lot of social impact, all right, but who never gave nanotechnology the least little chance of actually happening: It's really just science fiction, isn't it?

And at that point Roger Duncan had the reaction of many laymen when confronted with a new and controversial idea that they imagine is over their heads: "What do the experts think?"

Unlike the average layman, Duncan was extremely well positioned to find out. He was the president of a nonprofit foundation, Futuretrends, Inc., whose purpose was to monitor new technologies and educate the public about their possible effects. Separately, he knew that Susan Hadden had numerous grad students at her disposal, and so he offered to bankroll a special project evaluating the entire nanotech dream.

Thus it came about that in September of 1988 Eric Drexler flew down to Austin to give the inaugural lecture of a graduate-level course in which over the next two semesters sixteen students would read through a list of books and articles on nano topics, compile a roster of scientific experts qualified to venture an opinion about the subject, and conduct a survey as to what these experts thought of nanotechnology. At the end of it the students would tabulate the expert opinion and write up a report with their own findings, conclusions, and considered recommendations. The results would be published in 1989 by the University of Texas under the title *Assessing Molecular and Atomic Scale Technologies (MAST).*

As would be learned by anyone else who'd ever tried it, collecting "expert opinion" about nanotechnology was easier said than done. For one thing, the technology did not exist as yet, which meant there were really no "experts" in it to begin with. Besides which, the subject cut across so many disciplinary boundaries that anyone expert in one isolated part of it — chemistry, let's say — was not necessarily qualified to evaluate any other aspect — nanocomputer theory, for example. Then again, few scientists had even so much as heard of Eric Drexler or "nanotechnology," and among those who had, their knowledge of the subject was often of the book-review or gossip-column variety. And as for those few scientists who actually knew anything about the subject on a firsthand basis, well, most of them would sooner eat carpet tacks than tell you what they really thought of it, especially for the historical record.

This was doubly true when it came to saying what they thought of Eric Drexler personally.

"His name just automatically spawned negative feeling," recalled Lynda Cobb, one of the grad students involved in the project.

"It was kind of a weird situation. Some people didn't like him because they thought he was trying to glamorize the issue. Some people thought he was just a sci-fi weirdo. And then some people sounded like they actually knew his background and knew he didn't have a Ph.D., so he doesn't know what he's talking about."

Roger Duncan had similar recollections: "It was a very heated debate among the scientists. Some of them thought Drexler was absolutely right about everything; they had no problem with it. Others thought he was this egotist gathering followers. And they thought that he had no proof of any of it, so they didn't want to be associated with it in any way."

This presented a problem. These "experts" were from the best places — Caltech, MIT, Princeton, Purdue, Harvard, UCLA, Cornell, the National Science Foundation, and so on — and many of them had very definite views concerning nanotechnology, but in a major act of moral courage they refused to have their names printed next to their thoughts on the subject. In fact, they'd participate in the survey only on condition that their identities would never be revealed to anyone, forevermore and in perpetuity.

For the sake of getting on with their work, the students, faculty, and everyone else involved in the study took the required oath, and thereafter operated strictly under the code of *omertà.*

How strange, then, that when the "expert" opinion finally arrived it turned out to be heavily favorable toward nanotechnology! When the students asked, for example, "What do you see as the most troublesome bottleneck in the development of nanotechnology?" the answer was that there were really no "bottlenecks" worthy of the name.

"The notion of a bottleneck is troublesome," said one of the nameless experts. "When someone travels between the East Coast and West Coast and they run into a mountain, they go around it. Not only that, but if someone thinks something is extremely difficult, they may be surprised to find it is easy to do."

"Bottlenecks depend on the time scale," said another. "For a 100-year window, there are no real bottlenecks."

"All will require a great deal of research," another one said, "but I don't see one as causing the whole thing to come to a halt."

Worries about the uncertainty principle were mostly laid to rest. When the ghost experts were asked to rate the validity of the statement "Nanomachines would violate physical principles, especially the uncertainty principle," only 8 percent rated it "valid, possibly insurmountable," whereas more than 50 percent rated it flatly "not valid."

The *MAST* report ended with the inevitable recommendation that the government establish an advisory committee to oversee the development of atomic and molecular technologies — this to help ward off a molecular "disaster." More important than that, though, was the group's prediction that the interdisciplinary nature of nanotechnology would pose a special barrier to its realization. "We found that the most serious impediment to progress was likely to be the fact that members of the many diverse disciplines working at the nanoscale would not be aware of the rapid advances made outside their own fields."

This was a sharp insight that in time would prove to be entirely correct. Some chemists, for instance, were unaware of Hans Dehmelt's work with Penning traps, or of the atomic-manipulation abilities of STMs, for a long while after the fact.

In Drexler's eyes, the Texas study was the usual fiasco, another prime example of his views being distorted beyond all recognition. The package sent out by the students, he thought, gave a misleading, excessively futuristic cast to his overall theory, an impression he'd been given no chance to counterbalance, even when he'd volunteered to have them send out corrective information at his own expense.

As it was, even an accurate statement of his views yielded up a science-fictional picture of things. Drexler, after all, was preaching revolutionary, apocalyptic, semi-unbelievable notions. He talked about putting an end to aging, illness, and poverty; he talked about reviving people who had been in the deep freeze for years; he spoke about building earth-orbital-height skyscrapers and about traveling to the stars; he enumerated strategies for avoiding a worldwide molecular auto-da-fé at the hands of his tiny programmed assemblers.

How could anyone imagine that such a person was just not a

little off-beam? Even worse was the fact that preaching this stuff tended to attract those who decidedly were. There was the chap who came up to Eric at a conference — perfectly ordinary guy, absolutely normal in every way — and then started in on this complex analogy between nanotechnology and the Kingdom of God as described in the Bible, about *Engines of Creation* as the fulfillment of a prophecy, about Eric himself as having been sent here by God and the angels . . .

In the wake of this and other such occurrences there arose a concern on Chris and Eric's part about the issue of Eric's "followers," whose ranks were steadily growing in number. Drexler, indeed, was becoming something of a cult object: Captain Future, our fearless leader, the farseeing genius who'll drive the human race to its rightful destiny among the stars.

All of this came to a head at the Seattle "Nanocon" in February 1989, a major gathering of the nanoclan, first of its kind to be held on the West Coast. It had been organized by John Quel, a technical writer for Boeing Computer Services, and Jim Lewis, senior scientist at Oncogen, a genetic-engineering firm, both of them members of the Seattle-area NSG. Chris and Eric had wanted to hold further weekend retreats out west — they'd already attended one in a cabin on the Oregon coast in January 1988, and there'd be another in the Sierra south of Lake Tahoe in June 1989 — but they didn't say no to bigger assemblages even when, against their better judgment, one of them brought together both science and science fiction.

"Neither a fan gathering nor a true scientific conference," said John Quel, the conference chairman, "it shamelessly borrowed elements of both."

That seemed accurate. Held at the University Plaza Hotel near the University of Washington campus, the event featured "Guests," some of whom were scientists (John Cramer, physicist; Bruce Robinson and Ned Seeman, chemists; and Jim Lewis, biologist), others of whom were science-fiction authors (Marc Stiegler and Greg Bear), and one of whom, as an aid to confusion, was both a physicist *and* a science-fiction writer, Gregory Benford. Eric Drexler was "Guest of Honor" and star of the show, which was attended by eighty or so members of the nano faithful — "Eric's adoring fans," in the words of one cynic.

Steps were taken to keep matters on an even keel. "Discussion of exotic long-term consequences should remain the province of science fiction," Jim Lewis said at one point. "Interjecting such ideas into policy discussions scares people and distracts attention from the critical questions that need immediate attention."

Drexler himself went a bit further. At the beginning of the "Critical Path Panel" he asked Lynda Cobb, who was there from the University of Texas study, "to say a couple of words on that project."

Mainly she reported how hard it was to get the batch of experts together. "About half said that they would not participate because of the reputation that nanotechnology has associated with it. They called people involved with it 'the lunatic fringe.' I find it very interesting that the existence of a science-fiction-oriented group such as this is making it very hard for nanotechnology to be taken seriously in the technological arena. The paradox seems to be that if you want it to come about, you can't talk about it."

An excellent point. How did you make a revolution without seeming like a revolutionary? How did you advocate radical measures without coming off as a radical? Eric's answer was that you kept focused on the near-term and the technical. And you asked your adoring fans and followers to *please cool it!*

Which Drexler now proceeded to do.

"I would emphasize," he said, "that I have been invited to give talks at places like the physical sciences colloquium series at IBM's main research center, at Xerox PARC, and so forth, so these ideas are being taken seriously by serious technical people, but it is a mixed reaction. You want that reaction to be as positive as possible, so I plead with everyone to please keep the level of cultishness and bullshit down, and even to be rather restrained in talking about wild consequences, which are in fact true and technically defensible, because they don't sound that way. People need to have their thinking grow into longer-term consequences gradually; you don't begin there."

Keeping the cultishness and bullshit down. This now became a high-priority item among the nano faithful.

* * *

One thing that would help counteract the Captain Future image, Drexler knew, was getting his Ph.D. It would do absolute wonders for his credibility.

After several years at MIT, Eric had piled up enough academic credit and residency for a doctorate. All he needed was to pass the written and oral qualifying exams, write a thesis, and defend it. The question was, what department at MIT, or combination thereof, would approve a program of the type he was interested in? Molecular engineering didn't fall neatly under any single rubric: it wasn't chemistry, it wasn't biology, it wasn't physics. Nor was it computer science, engineering, or any other separate thing; rather it was a mix of all the above.

MIT allowed for "interdepartmental" doctoral degrees, but they were highly unusual specimens and not particularly encouraged by the faculty. The school put you through all sorts of administrative hoops and tortures before they'd even let you get started, much less award the degree. You'd have to carve out some interdisciplinary field for yourself, assemble a committee from the various relevant departments, write out a proposal, and then get the whole shebang approved by both your "home" department plus the dean of the graduate school. Only then could you begin actual work.

But in truth such an arrangement was highly agreeable to Eric Drexler. Not only did he have what might be called a strong aesthetic preference for forging his own path, in the instant case he seemed to have no other choice. The document he intended submitting as a thesis, after all, was the technical text he'd started taking notes for while teaching the Stanford course, in preparation for which he'd put together materials of the most amazing variety. He'd amassed countless equations, numerical calculations, scaling laws, elastic-deformation-rate curves, diagrams of various atomically precise structures, and so on and so forth. This was the true hard-core stuff, precisely the nuts-and-bolts minutiae that would convince his peers — if, indeed, anything would — that he was describing a real prospect and not just "science fiction." Why not get his degree by writing a thesis that could also be published as a book — earning his union card and presenting his vision to the technical world — all in one fell swoop?

On one of his visits back east — this was in 1989 — Eric stopped in at MIT and told Chris Fry, the NSG member who was closest to Marvin Minsky, that he was getting serious about going for his degree. Fry later mentioned this to Minsky, who thought it was a wonderful idea.

Marvin Minsky, it must be said, was tailor-made to supervise the project. For one thing, he'd always been quite the iconoclast: he seemed to go out of his way to adopt heretical positions.

"Nuclear explosions aren't so terrifying, because they're not self-replicating," he once told a nano audience. "They're just *irritating.*"

Minsky, like Hans Dehmelt, took a distinctly revisionist stance when it came to the meaning of quantum mechanics. The message of quantum mechanics, he said, was not the customary platitude that "things at the atomic level are uncertain" — just the opposite. The truth was that it was in classical mechanics where things were uncertain, at least insofar as atoms were concerned: the atom was inexplicable under classical laws, according to which the electron radiated away its energy and spiraled into the nucleus immediately.

In quantum mechanics, by contrast, an atom was either in one definite quantum state or it was in some other definite quantum state. How could you get more certain than that?

"It should be on the front page of the *Times*, or somewhere, that quantum mechanical atoms are certain!" Marvin Minsky said. "They're more certain than anything else we know of!"

Then, too, Minsky had always been one of Eric's biggest supporters at MIT. He'd written the foreword to *Engines of Creation*, where he'd pronounced the book to be "enormously original," "ambitious and imaginative," and "technically sound." And he had nothing but praise for Drexler as a person. "The great thing about MIT," Minsky once said, "is that you have great teachers. *I had Eric Drexler.* Every time I ran into him I learned a new field in about thirty seconds."

So who better than Marvin Minsky to supervise Eric's doctoral work?

Not that it helped. The hope was that Drexler's proposal, his publication record, his two previous MIT degrees, his patents (he'd patented his solar sail and his space-based method for making thin films), plus the weight of the great Minsky himself, a formidable presence even at his home institution, all this would combine to get Eric through a program in Minsky's own department, electrical engineering and computer science, or EECS. Accordingly, Eric applied and was admitted to the EECS graduate school for the specific purpose of pursuing an interdepartmental Ph.D. degree.

At which point all hell broke loose.

"Things came unhinged when Drexler didn't want to take any exams," said a member of the department's graduate committee.

More precisely, he didn't want to take the exams that were normally required of an EECS degree candidate. Drexler's reasoning was that he wasn't going for an EECS degree, he was going for an interdepartmental degree, so why should he take the usual exams?

But exams were only the tip of the iceberg.

"He had a very unusual background for someone in electrical engineering and computer science," said a faculty member in the department. Indeed, Eric's bachelor's in interdisciplinary science (a degree that MIT offered for only ten years or so before discontinuing it), and his master's in aeronautics and astronautics were not logically related to EECS in any obvious way. Besides which, he'd never taken even so much as a single course in electrical engineering or computer science, the very department he was now asking to be his institutional "home."

But to Drexler himself, all that stuff was totally irrelevant. EECS would only be acting as the host department, the institute's paper-shuffler-of-record. It didn't matter that he hadn't taken any EECS courses because he wouldn't be getting an EECS degree, he'd be getting an interdisciplinary degree.

The final straw, however, was Drexler's thesis proposal itself. It was a wee bit too cutting-edge for the department's taste.

EECS, as some of its faculty members freely, almost proudly, told you, was not one of the more forward-looking divisions at MIT. It was, after all, an engineering department. Engineering departments could not afford to be out there on the cutting edge, pushing the boundaries and experimenting with all kinds of new and unproven designs.

"You understand the problem," said Gerald Jay Sussman, an electrical-engineering professor who supported Drexler. "You build a bridge, it falls down, somebody's going to get angry. The idea is that you don't want it to be one of your students who designed that bridge. As a consequence, engineering departments tend to be more conservative about their endorsements of plans of research than nonengineering departments. When they give out a degree, they want it to mean something. You're certifying someone, and the certification part is what people were worried about."

So in all reason what could EECS be expected to do when confronted with a Ph.D. proposal that described a realm of molecular machines and molecular computers, mechanisms that not only did not yet exist, but whose chances of ever doing so were exceedingly dim in the view of some of the more vocal members of the graduate committee? Who'd have expected them, an engineering department, to approve of such a thing?

"People on the committee were very, very nervous about this," said one of the committee members.

A meeting was scheduled for October 2, 1989, at which Drexler's case would come before the full and formal EECS Committee on Graduate Students. Nine voting members attended; in addition, Minsky and Sussman, who were not on the committee but who were EECS professors, came in as friends of the court, as it were, to plead Drexler's case.

A battle of historic proportions erupted — and Drexler's side lost.

"Although no specific vote was taken, the committee was reluctant to approve an interdisciplinary doctoral committee for Drexler," one of the members said later. "This has been our response to such requests for many years; unless there is a compelling reason to establish such a committee, we strongly prefer to have our doctoral students follow our standard exam procedures."

And that was the last EECS ever saw of Eric Drexler.

* * *

In California, meanwhile, better progress was being made on a different educational front. Todd Gustavson, an eighteen-year-old high-school student, was building his own scanning tunneling microscope — the kind of instrument with which people at the fanciest research centers were now moving atoms — and he was doing it at home, in his father's workshop, for a total cash outlay of about two hundred dollars.

Todd had gotten the idea in January 1987, back when he'd attended a lecture by Cal Quate, the Stanford physicist who five years earlier had made an en route detour from London to Zurich to learn all about STMs. In the years since, Quate had worked with STMs of his own at Stanford and had become a genuine evangelist on the subject.

Todd and his father, David, a physicist at the Stanford Linear Accelerator Center, became equally wrapped up as they listened to Quate describe the device and what it could do. There was really not all that much to an STM, Quate said, just a needle tip, sensing

Todd Gustavson and his homebuilt scanning tunneling microscope. (*David Gustavson*)

circuitry, and a means for precisely positioning the sample. But with it you could move atoms.

"It suddenly seemed pretty simple," Todd recalled. "We said, 'Gee, we could build one ourselves. It looks like fun.' "

Todd's father was an electronics hobbyist, which in strictly operational terms meant *parts bins* — rank upon rank of these little plastic sliding trays crammed with supplies and components of every description: banana plugs, alligator clips, wires, leads, resistors, capacitors, condensers, rectifiers, transistors, Wheatstone bridges. There were soldering irons and circuit boards all over the place, plus testing and monitoring equipment: oscilloscopes, circuit testers, and absolutely endless varieties of "meters." You could make anything short of an atom bomb from out of those parts bins.

There was only one thing Todd Gustavson lacked when it came to building an STM: a "piezo." Every STM had to have a "piezo."

Piezoelectric elements, formally speaking, were these little miracle items, pressure-sensitive crystals that generated electrical currents in response to applied mechanical forces, or, conversely, generated mechanical forces in response to applied voltages. You could feed in some electrical current to a piezo, and it would reward you by expanding or bending just ever so slightly. A piezoelectric element was the part of an STM that controlled the movement of the needle tip: the needle was attached to the piezo, which by the application of the appropriate voltages could be made to deflect simultaneously in the X, Y, and Z directions (forward and backward, side to side, up and down), all under precise control.

When Todd Gustavson attended Cal Quate's lecture, one of the things he came away with was his very own piezo. There were some extras lying around in the lab, and a grad student who worked with Quate, Doug Smith, just handed him one.

Todd went on to build the world-record bargain-basement STM. He cut every conceivable corner. Back then he had braces on his teeth and so to attach the tungsten needle to the piezo he put two dabs of solder on the piezo, filed a V-shaped groove across the solder dabs, placed the needle into the grooves, and held it in place with rubber bands that came with the braces. "This method of securing the needle," he noted, "is very simple compared to techniques used in other STMs, which use a tiny machined block of brass and a set screw to hold the needle in place."

Vibration damping, similarly, was a matter of placing some cheap shock-absorbing materials between the STM and the floor. Where Binnig and Rohrer had suspended their STM on springs inside of a stainless-steel frame that in turn rested on a complex system of copper plates and magnets — truly the high-rent approach — Todd started out with a wheelbarrow inner tube, laid a concrete block on top of it, and then piled up a stack of alternating plastic plates and rubber tubes. The STM sat at the top, like a Buddha. It was a little wobbly, this vibration-damping tower — it resembled a step pyramid — but anyway it worked.

The chief mechanical hurdle was getting the needle to come into close contact with the sample without actually crashing into it — something that happened all too often even in the best STMs, even with all their complicated gearing and fine-tooled machinery. In place of the latter Todd used the voice coil from a loudspeaker. Like piezos, voice coils moved in response to electrical currents: "Very fine control of the current through the speaker coil moves the attached sample as little as an angstrom." That was about the diameter of an atom.

Rough-and-ready as it all was, the device worked well enough. (Despite his vibration-damping pyramid, the STM picked up his mother's footsteps as she walked from room to room.) It was Todd's hope, therefore, to be able to see individual atoms. He took a small piece of graphite, whose hexagonal-shaped atomic patterns would be easily recognizable, and placed it into the microscope.

"We got to the point of seeing a lot of surface features of the graphite," he said later. "At one point it really looked like there were atoms on the screen, but it was never reproduced consistently enough to be sure."

In April 1989, after working on the machine for a year and a half, Todd entered his homebuilt STM in the Santa Clara Valley Science and Engineering Fair, where it won the grand prize. His paper, "Design of a Simplified Scanning Tunneling Microscope," took first place in the technical-paper category. Later he entered it in other competitions, racking up more than twenty-five awards in the space of a few months.

One of them was a free trip to Stockholm to watch the 1989 Nobel Prize ceremonies. He stayed at the Grand Hotel, same as the Nobel laureates. He even saw Hans Dehmelt accept his award — he of the pet particles Astrid and Priscilla.

Todd, who lived in Los Altos, only about a mile or so from Chris and Eric, was fully aware of the possible nanotech applications of his device.

"STMs," he'd said in his technical paper, "could someday be instrumental in examining or creating custom-made structures atom by atom."

And this was surely food for thought — homebuilt STM hobbyists creating custom-made structures atom by atom.

"That would be a second-order sort of problem, one which I never got to thinking about, or worrying about," Todd Gustavson said. "It's not likely that it would be very easy. You'd have to have a lot of control."

Nevertheless, he didn't say it couldn't be done.

Which meant that anyone who was sufficiently paranoid could now begin to imagine scenarios where some teenage atom-hacker was home alone — out in the garage or upstairs in the playroom — spelling his or her name out in atoms. Was it such a big step from there to joining a few random atoms together, producing a molecular wheel or axle? And from there to automating the process, programming the family Macintosh so as to keep the atomic-construction routines going overnight?

(. . . patterned molecular arrays taking shape . . . wheels and axles proliferating like tadpoles . . . gears meshing . . . bearings turning . . .)

So that next morning, when youthful atom-crafter wakes up and takes a look at the STM screen, what should appear there but . . . molecular robot arms all over the place! Oodles of them! Scads of them! Out-of-control nano machinery right there in the rec room!

"It was easy! I just used some of Eric Drexler's really neat pictures as models!"

Was it really so improbable? Was it really such an absurd fantasy? Would anyone claim that such a thing couldn't ever possibly happen out there in teenage hacker land?

Nanotechnology was a new metaphysics, a whole new way of seeing the world. Nature was suddenly much more pliant, yielding, malleable. No longer would you have to take brute matter as it came — a fortunate circumstance in view of how *stupid* ordinary matter was thought to be by some of the more conceptually advanced nano-advocates.

"If you look around you right now you see all these things built out of dumb materials, this material that is eternally stupid," Mark Miller said. "This table is *so stupid,*" he said, giving his coffee table a kick. "There are enough atoms in that table that if they were being used effectively they could be generating an amount of computing power that exceeds the total amount of computing power in the world right now. *All* the stuff we're currently surrounded with could be doing so much more for us than it is. One way to understand nanotechnology is that suddenly material is so malleable that in some sense it's all software."

What this meant in practical terms was demonstrated by JoSH Hall, of sci.nanotech fame, who one morning while driving in to his job at Rutgers came up with a fabulous new nano-invention, the Utility Fog.

His seat belt had been bothering him — it was binding, chafing, cutting him in the shoulder.

"This is really dumb!" he thought. "This wouldn't happen in a nanotech world. With nanotechnology you'd have something a lot more sophisticated than a seat belt!"

"The obvious thing is to be totally encased in a form-fitting enclosure," he said later. "You know, like when you transport fragile equipment they have these containers that are filled completely with foam, with form-fitting cavities that you put the equipment in. What you'd want for a nanotech seat belt would be something like that, something that in the event of an accident would act like this foam, but otherwise would act like the air in the car."

That was the basic concept of the Utility Fog: an invisible, form-fitting cloud of matter. It would look and feel like air, but under special conditions, such as in an accident, it would take on the aspect of a physical object, like a block of rubber, for example.

You could create such an entity, he thought, by breaking up a quantity of matter into tiny invisible bits — pieces that were larger than atoms but still too small to be visible — and then dispersing them out in all three dimensions. Then all you'd have to do was to link those bits and pieces together: you could put tiny hooks or arms on them, for example, so that each of them could connect up with all of its immediate neighbors. And of course you'd have to make them individually and collectively controllable, so that they'd be able to act in concert to achieve a consistent effect.

JoSH visualized an array of microscopic robots each about the size of a human cell, with lots of arms and hooks sticking out every which way. It would be an interlocking system, a connected, functional network, like a sponge, an invisible sponge of tiny robots — "foglets," as he called them.

Right away, though, he saw some problems. How could you move inside of an organized meshwork of foglets? How could you even breathe?

But then again people breathed lots of such stuff already, and they didn't even know it. Air was filled with some of the most vile and lurid robotic monsters imaginable: microorganisms, bacteria, viruses, protein molecules, pollen grains, yeasts, spores, funguses, rusts, tiny bugs. People breathed in all of that stuff every minute of every day and they never even gave it a thought. Same with his tiny mechanical robots: you'd breathe them in and out constantly and never be aware of it; they wouldn't do you any harm.

Motion through the Utility Fog presented more of a problem, but one that was easy enough to solve so long as the foglets were under computer control. They could be programmed to unlink themselves in front of you as you approached and then close ranks again as you moved through. That way, you'd pass through the foglets like a slide on a zipper; you'd move right through the stuff, just like it wasn't there. You could fill a whole room with Utility Fog and you wouldn't even notice it.

You could do some really neat things with the Utility Fog, he decided. You could have the foglets bunch together and disport themselves in any arbitrary fashion. For example, you could have them assume the physical shape, form, and function of a bottle, and it would even work like one: you could pour liquids in and out. You could call up any physically allowable object; you could sheerly bring it into existence, out of thin air, on command.

"It will be almost any common household object, appearing from nowhere when you need it (and disappearing when you don't). Power tools, kitchen implements, and cleaning chemicals would not normally exist: they or their analogues would be called into existence when needed and vanish instead of having to be cleaned and put away. It can act as shelter, clothing, telephone, computer, and automobile. The same Fog that was your clothing becomes your bathwater, and then your bed. Virtually every physical task or job humans do could be performed by the Utility Fog."

You could make a whole house out of the Utility Fog, he thought.

Or even a whole city . . .

"The Fog can act as a generalized infrastructure for society at large. Fog City need have no permanent buildings of concrete, no roads of asphalt, no cars, trucks, or buses. It can look like a park or a forest, or if the population was sufficiently whimsical, ancient Rome one day and Emerald City the next."

That's how malleable reality would be, that's what matter-as-software would be like . . . after the nanotech breakthrough.

Finally, even the mainstream media began to pay some attention. In December 1988, *Fortune* magazine ran a story, "Where the Next Fortunes Will Be Made," which included several paragraphs about making money from nanotechnology.

"For a truly spectacular look ahead," the story said, "consider the entrepreneurial possibilities inherent in molecular-scale machines capable of extracting gold from seawater or diamonds from egg yolks. This is nanotechnology, the 21st century's answer to alchemy."

Current manufacturing and engineering techniques, it said, would become "obsolete." Who, then, would be making any money? "Software designers who will write the programs to tell the nanomachines what to produce."

The following month, in January 1989, *Omni* magazine ran a full-length interview with Drexler — "Mr. Nanotechnology."

And then, in the summer of 1990, the *Whole Earth Review* published what was, at the time, the first and only attempt at an extended and serious criticism of the subject.

"Critique of Nanotechnology" was written by Simson Garfinkel, a former MIT chemistry major who was now a reporter for the *Christian Science Monitor.* Garfinkel had gone out and interviewed some working scientists — all of them MIT faculty members or grad students — about Eric Drexler and his grand creation.

The working scientists, it seemed, were not impressed.

"Drexler discusses these molecular systems as mechanical systems," said Robert Silbey, an MIT chemistry professor. "He bangs them and they go." The problem, in Silbey's view, was that "molecules are not rigid — they vibrate, they have bending motions."

("This is *news?*" Eric thought.)

Rick Danheiser, another chemist, was quoted as saying: "I see some antiaromatic structures that can't possibly exist. It's unfortunate that he draws something that doesn't look so good, because a lot of people see it and discredit the whole thing."

"My main criticism of nanotechnology, or more in particular, of Drexler," said James Nowick, a grad student in organic chemistry, "is that he's coming forth as being sort of a visionary without actually doing anything. . . . Whatever he is putting forth as science has to be tempered by the fact that we are dealing with somebody who is basically making predictions. In my field, if you have a prediction of how something will work you can't just go publish that. You really have to have scientific results."

Garfinkel's piece was written in a mock-hypertext format, so that Drexler could respond to his critics, with his answers immediately following their complaints.

To Silbey's observation that molecules vibrate and bend, Drexler's response was: "Every physical object is a collection of atoms; nanomachines will simply be very small physical objects. Everything vibrates, everything bends, and machines work regardless. Single-atom gear teeth will indeed bend under load (why would anyone assume that I think otherwise?), but they will also turn the gear, given any sort of reasonable bearing."

To Nowick's claim that nanotechnology was a bunch of predictions, Drexler said: "Nowick is correct that predictions are not publishable in many fields of science. However, nanotechnology is not a branch of science (as I have taken pains to point out in *Engines of Creation*); it is an engineering discipline based on established science. Engineering projects are often discussed and written about before they are undertaken."

And then there was Rick Danheiser's quote about "some antiaromatic structures that can't possibly exist." Drexler had given an answer to this, a long technical reply, but much more interesting was what Danheiser himself had to say about his own quote after seeing it published in the *Whole Earth Review.*

"I never made such a statement," said Danheiser. "I would never make such a statement! It's nonsensical. It's a misquote: that 'interview' was conducted in the MIT cafeteria while we both ate lunch.

" 'Antiaromatic' systems do exist; examples have been known for a long time. I devote half a lecture in my introductory organic chemistry course to this subject. What is in question is not the ability of antiaromatic compounds to exist, it's their stability relative to other possible 'building blocks' for nanotechnology that I was commenting on. In the context in which this came up in my conversation with Garfinkel, I was in fact defending Drexler! In short I was explaining that criticizing Eric's designs because they incorporate antiaromatic structures misses the point of what he is saying."

The *Whole Earth Review* piece, nevertheless, had introduced its criticisms with the heading: "Chemistry says it can't happen."

13

"Are Molecules Sacred?"

If its first couple of years were any indication, the 1990s would be the nanotechnology decade. During this time laboratory researchers spelled out the IBM logo in atoms; invented the world's first one-atom switch; created the world's first self-replicating molecule; constructed a molecular "shuttle" and "train"; and performed several other stunts, tricks, and minor molecular miracles. *Science* published a special issue devoted to nanotechnology ("Engineering a Small World: From Atomic Manipulation to Microfabrication"), and *Nature* sponsored a nanotechnology conference in Japan. A new scholarly journal, *Nanotechnology*, started up in England, while in the United States, MIT Press published *Nanotechnology: Research and Perspectives*, the proceedings of Drexler's own Foresight Institute conference on the subject.

Scanning tunneling microscopes — for which Binning and Rohrer had now won the 1986 Nobel Prize for physics — were soon accompanied by a glut of other so-called proximal probes, instruments designed either to picture or actually manipulate atoms and molecules. There were atomic-force microscopes (AFMs), friction-force microscopes, magnetic-force microscopes, and electric-force microscopes; there were scanning thermal, optical absorption, scanning ion-conductance microscopes, and many more besides.

Digital Instruments, of Santa Barbara, California, was selling its own brand of STM, the NanoScope, by mail order. ("Easy to learn and use," said a company brochure. "No special skills are required. Initial set up takes just a few hours and the space of a desk top; you can produce images of atoms on the first day.")

Start-up companies such as Protein Design were doing R&D on synthetic proteins, while new journals such as *Protein Engineering* devoted themselves entirely to a subject that ten years earlier had still seemed like a far-off dream. Even the professionally glum *Limits to Growth* people, returning in 1992 with a sequel, *Beyond the Limits,* conceded that molecular engineering actually might have some prospects, might be able to help stave off their still-projected "global collapse."

"Breakthroughs in nanotechnology and biotechnology," they wrote, "are beginning to allow industry to carry out chemical reactions the way nature does, by careful fitting of molecule to molecule."

Suddenly, all the outward appearances were that nanotechnology was the wave of the future, and just about ready to crest.

Drexler himself had unparalleled successes in the new decade. He'd get his doctorate from MIT; publish his technical text, *Nanosystems;* testify before Al Gore's Senate subcommittee; and give nanotechnology briefings at the White House Science Office and at the Pentagon. His creation was enshrined by the Xerox Corporation, whose Palo Alto Research Center, the famous "PARC," underwrote a "computational nanotechnology project" exclusively for the purpose of doing computer simulations of Drexler's gears, bearings, and associated molecular gadgets.

There were, true enough, one or two wrinkles in the fabric of perfection. In its nanotechnology issue, for example, *Science* had said that Drexler was "anathema" to some researchers, and quoted Phillip Barth, of the Hewlett-Packard Company, as saying of Drexler: "The man is a flake." (In a letter to the editor published in a subsequent issue, Drexler complained that this was an example of how some people, especially "silicon micromachine researchers, seeking to dismiss molecular nanomachines, resort to mere name-calling.") The *New York Times,* for its part, quoted Cal Quate as saying of Drexler: "I don't think he should be taken seriously. He's too far out."

But such carpings were entirely predictable: paradigm shifts — where the orthodox scientific worldview was under attack by an unsightly new version of events — were regularly accompanied by major amounts of henpecking, ridicule, and flagrant gnashing of teeth. And in fact such behavior wasn't confined to the sciences. It was familiar in various disciplines down through the ages, as people heralding vast new changes were *always* dismissed by the reigning establishment; it was practically a law of nature. At one of Drexler's Foresight conferences Charlie Coulter, of the National Institutes of Health, read out to the assembled multitude a weirdly fitting passage from Machiavelli: "There is nothing more difficult to take in hand, more perilous to conduct, or more uncertain in its success than to take the lead in the introduction of a new order of things, because the innovator has for enemies all those who have done well under the old condition, and lukewarm defenders in those who may do well under the new."

In the front row of seats, Eric Drexler listened to this, raised his eyebrows, and nodded his head.

If there was an emblematic signpost of the "new order of things," it was the IBM atomic logo that had been revealed to the world in the April 5, 1990, issue of *Nature.* Two researchers at the IBM Almaden Research Center, Donald Eigler and Erhard Schweizer, had used a scanning tunneling microscope to spell out "IBM" with thirty-five individual atoms of xenon. They'd pushed the atoms around on a nickel surface, like a bunch of checkers, until they were arranged in the shape of the three separate characters, nine atoms for the letter *I,* thirteen for *B,* and thirteen more for *M.*

Like everything else about the STM, the process of using it to form the atomic letters wasn't particularly complicated or difficult. The researchers had started out with a clean surface in an ultrahigh vacuum into which they'd introduced a small quantity of xenon. It was then a matter of locating a random xenon atom — they showed up as bumps, 1.6 angstroms high, on the surface — and physically dragging it with the STM tip ("at a speed of 4 Å per second") until it was in the desired position, and then repeating the process for each successive atom.

To minimize thermal energy, kT, the procedure took place at a temperature of 4 degrees Kelvin — close to absolute zero. "This notwithstanding," the researchers said, "it is important to realize that the process we describe for sliding atoms on a surface is fundamentally temperature independent."

It took them about an hour to form each individual letter. Nevertheless, the experimenters discovered that they could work with a given atom just as long as they wanted to: "The stability of the sample and of the microscope due to the low temperature are such that one may perform experiments on a single atom for days at a time."

Experiment on a single atom for days at a time. It was the kind of claim that physicists would have laughed out of court, would have regarded as the worst sort of insane crackpottery, right up until the moment it happened.

The IBM logo, of course, inspired all sorts of copycat feats as people in applied-physics labs the world over now used their own STMs, AFMs, and other proximal probes to write out their own name or initials, their lab's name or acronym, Einstein's equation $E = mc^2$, and Specially Meaningful Words (such as *Peace*) — all these things were now lovingly depicted in atoms or small bunches of molecules.

Atoms, in fact, became the preferred medium for nano-artistic expression as working scientists used STMs to draw some of the dumbest and silliest pictures ever recorded — all of them, naturally, in the name of "research." A famous portrait of Albert Einstein, the one with his tongue sticking out, was reproduced at the nanometer scale, "on the surface of a mixed-ionic conductor"; a map of the Western Hemisphere was drawn with tiny blobs of gold at the scale 1:10,000,000,000 ("STM opens up a new world," said the punning caption); an atomic gingerbread man — a little cookie-cutter "molecular man" — was rendered on a platinum surface with twenty-eight individual molecules of carbon monoxide. What a gas!

World's dumbest picture! Stupidest sculpture! Smallest map!

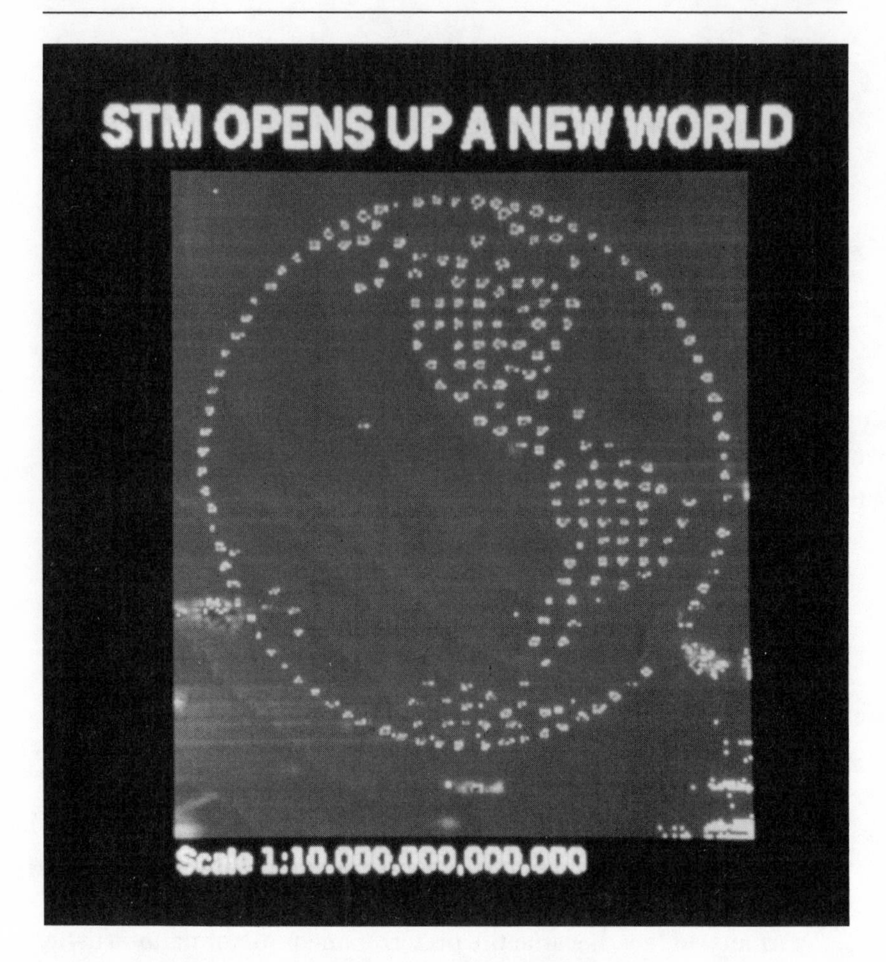

Map of the Western Hemisphere,
reduced by a factor of ten trillion. (*IBM Corporation*)

These were dangers that not even Eric Drexler had been able to warn against.

About the only one not impressed by any of these atomic high jinks — especially not by the IBM thirty-five-atom logo — was Hans Dehmelt.

"All this business of these big claims by IBM, what they have there is a piece of crystal with *pimples* on it!" he said. "They don't have single atoms! I mean, these atoms are not identities, they're exchanging their guts all the time with the others."

Even at 4 degrees Kelvin?

"Well, excuse me, the electrons are exchanging all the time! And the nuclei even jump around. These *surface pimples* are jumping around all the time! At low temperatures it probably takes a long time, but it still happens."

That was all the vaunted IBM atomic logo amounted to in Hans Dehmelt's eyes: *surface pimples!*

How truly welcome it was, then, when some of those bold atomic explorers actually deigned to name an actual purpose or two, a *use*, for one of their prize creations. IBM's Eigler and Schweizer, for example, now claimed that "it should be possible to assemble or modify certain molecules in this way. We can build novel structures that would otherwise be unobtainable. The prospect of atomic-scale logic circuits and other devices is a little less remote."

The following year, those "circuits and other devices" were less remote still, for by then Eigler and his colleagues had actually produced one such, "an atomic switch." It was pictured in full color on the cover of *Nature*, August 15, 1991.

The experimental setup was the same as before, a nickel surface in high vacuum and cooled to 4 degrees Kelvin, then covered with a layer of xenon. By placing the STM tip above a chosen xenon atom and varying the tip voltage, it was possible to make the atom oscillate back and forth between the tip and the surface. It was, in Eigler's words, "an atomic-scale bistable electronic switch."

It was also the first man-made, real-world prototype of what Drexler had envisioned years before, in 1977, when he'd come up with the idea of an atomic-scale computer. Even Cal Quate, who regarded Eric as "far-out," suddenly threw all caution to the winds.

"The prospect opened by this remarkable new body of work is the construction of electronic devices with atomic dimensions," he wrote in *Nature*. "A cluster of 1,000 atoms could represent one bit of information, in which case the entire contents of the Library of Congress, equaling 200 terabits, could be stored on a silicon disk 12 inches in diameter. (In contrast, it would probably take 250,000 compact disks of a similar size to store this information with current techniques.)"

That was the good news. The bad news was the length of time it would take to read and write that information. Quate calculated that even if you could write to the disk at the "exceptional" rate of ten million bits per second, "it would take 230 days to fill the disk. Worse still, it would take an additional 230 days to read the information."

Which sounded bad at first hearing. But on the other hand how many people could read the entire contents of the Library of Congress in 230 days? And who in their right mind would even want to?

"The solution," Quate said, "will be to work towards massively parallel reading and writing systems."

Putting the emerging nano-zeitgeist into words was John A. Armstrong, IBM Chief Scientist and Vice President for Science and Technology, who said: "I believe that nanoscience and nanotechnology will be central to the next epoch of the information age, and will be as revolutionary as science and technology at the micron scale have been since the early 70s. Indeed, we will have the ability to make electronic and mechanical devices atom-by-atom when that is appropriate to the job at hand."

And why not? The one-atom switch, after all, was here.

When Eric was turned down by EECS, practically the first thought that popped into Marvin Minsky's head was, "Well then, why not do it through the Media Lab?" Minsky wore two hats at MIT: not only was he the Donner Professor of Science in EECS, he was also the Toshiba Professor of Media Arts and Sciences in the Media Lab.

"I decided it was hopeless fussing with the EE department," he said afterward. "Some of them were annoyed at Eric's proposal for reasons I couldn't fathom. It seemed to me it was a paradigm-shift problem of some sort. I was really quite astonished about it."

Not so astonished was Drexler himself. The EECS graduate committee, after all, was known to harbor within its ranks a couple of extremely dedicated and one-track micromachinery researchers — people working on the "three little gears" sort of thing. One of them had told Eric that his nano stuff was just "conversational science" — all talk and no action. The other researcher was even less flattering: "He thinks that all this is garbage, for reasons I have never had explained to me," Eric said later.

Truth be told, there was never any love lost between the micro and nano factions of the miniaturization fraternity. At one of his Foresight conferences (among friends and supporters), Drexler had even said that micromachinery was "*irrelevant* to nanotechnology. It could stop dead today and it would make no difference to the development of nanotechnology."

Drexler's own version of what had gone wrong at EECS was that the micro people were worried about nanotechnology's making their own discipline obsolete.

"That's reading into it," said Jerry Sussman, one of Eric's supporters at EECS. "I think that that's a possible conjecture, but the fact of the matter is that people are much more complicated than that and there's a totally other possibility, which is that people who objected to his work were people who just felt that it was flakier than they liked.

"Drexler's arrogant and a bit obnoxious," he added. "On the other hand he's smart. He doesn't go through the standard paths people take, and a lot of standard people don't like people who don't take the standard paths."

People at the Media Lab, to put it mildly, were not so fussy about standard paths.

"The Media Lab is certainly not a conservative place," said Sussman. "In fact, probably the Media Lab exists because people considered other departments that they were in to be too conservative for some of the research they wanted to do."

Anyway, Eric revised his extremely nonstandard thesis proposal — it described, after all, a whole new technology — and took it over to the Media Lab.

By any measure, this was something of a comedown. In the first place, the Media Lab was not formally a part of the engineering school at MIT; it was based instead over at the School of Architecture and Planning, which would be an awfully strange place for a degree in "molecular nanotechnology," of all things, to emerge from.

Separately there was the fact that the Media Lab enjoyed a

less-than-sterling reputation even within MIT itself. The place was presided over by Nicholas Negroponte, an architect who early on in his career had conceived the architecturally trailblazing notion that there was no special reason why city-planning work had to be done by humans. It might just as well be done by animals, for example by a bunch of gerbils running through a maze: What was a city, after all, but a vast mazework of streets and buildings?

So Negroponte constructed a toy city out of stacked one-inch cubes — these were the buildings — and let loose into it a colony of gerbils. The idea was that the gerbils would run around the city like animals, toppling the play buildings and arranging the cubes in ways that were more convenient and personally satisfying to them, thereby setting new standards for contemporary architecture and city planning.

Well. An item like this could have no better home than a major metropolitan museum, and in 1970 the Jewish Museum, in New York, placed Negroponte's "architecture machine" on display, causing a local sensation.

It was precisely the type of venture that, because of its "creativity," would be highly regarded at the Media Lab.

"Most view the Lab as a lightweight," said Fred Hapgood in *Up the Infinite Corridor*, his book about MIT. "Something of a theme park, not involved with problems worth a grown-up's attention. . . . Tomorrowland."

A theme park! Tomorrowland!

Exactly the Captain Future image Drexler was trying to escape. But what other choice did he have? He wasn't about to start over at Stanford, Caltech, or wherever else.

Naturally, since the first thing people thought of when they heard the word "media" was television, Drexler was careful to include in his revised Ph.D. proposal some of nanotechnology's more remarkable information-handling and display applications.

"If what I have been arguing to technical audiences for the last several years is true," he wrote in the new proposal, "then the media systems of the future will eventually be based on the products of nanotechnology — that is, they will exploit hardware such as molecular tape memory systems able to store 10^9 terabytes of data per cubic centimeter; inexpensive computers with a computational capacity up to 10^{12} times greater than that of today's machines; flat-screen, full-color, high-resolution, 3-D displays; and so forth."

In the same document, he also threw down the gauntlet to his foes at EECS.

"Any critic wishing to save MIT from involvement in research on molecular nanotechnology," he wrote, "is hereby challenged to provide a written statement describing some substantial error in my analysis or conclusions and to defend that statement in a public debate. Barring this, I ask that this proposal be approved and that the critics be invited to my thesis defense."

Although it was IBM's atomic structures that collected the headlines and got pictured in *Time, Newsweek,* and so on, equally amazing molecular structures were being produced in traditional solution chemistry.

Solution chemistry was the old-style of manipulation of matter that Drexler hoped largely to replace with his own new brand of matter handling: mechanochemistry. Traditional chemists worked with bulk quantities of matter in test tubes, flasks, beakers, and so on, and induced chemical transformations through heating, stirring, and other large-scale processes. Drexler's mechanochemistry, by contrast, would proceed with atomic-scale precision by means of the individual, mechanical placement of molecules. An assembler grabbing on to a given molecule, transporting it to a reaction site, and guiding it forcibly into position — this would be mechanochemistry, with which, obviously, some spectacular feats of creation could be performed.

But then again some pretty spectacular feats were still arising out of the old-time "solution-phase" procedures. In 1990, for example, Julius Rebek created the world's first self-replicating molecule, a simple two-part structure. When the molecule split apart, the two halves separated off and chemically attracted new mates, forming two new molecules, each of which was an exact duplicate of the original. The reaction then began again, the two parts separating off, finding new counterparts, and so on. Once they got going, Rebek's replicators duplicated themselves millions of times per second.

Even Drexler was impressed.

"When I met with Rebek I told him that he is one of the chemists for whom I have the greatest respect, because he builds *systems.* But if you look at the size and complexity of that system, it's very slight."

Others were building systems of somewhat greater size and complexity. J. Fraser Stoddart, the British chemist, used traditional solution techniques to create molecules that approximated macro-scale mechanical systems.

First there was the molecular "shuttle." Stoddart meant this literally: he constructed a shuttle in the sense of a bus or a ferry, something which, as he said, went "back and forth along a specified route or path at regular intervals."

Implementing such a thing at the level of molecules involved synthesizing the four basic parts: the "track" along which motion would take place; the "stations" at which the shuttle would stop and reverse direction; bumpers or stoppers at the far ends of the track to prevent the shuttle from going too far and falling off; and the molecular shuttle vehicle itself.

Stoddart had done all this by the end of 1990, creating components that assembled themselves from solution and then proceeded to behave much like their Big World counterparts. At room temperature, the molecular shuttle moved back and forth along the track, like a bead on an abacus, at the rate of five hundred cycles per second.

Moreover, Stoddart actually saw some "uses" for such a device: "The molecular shuttle is the prototype for the construction of more intricate molecular assemblies where the components will be designed to receive, store, transfer, and transmit information in a highly controllable manner, following their spontaneous self-assembly at the supramolecular level. Increasingly, we can look forward to a 'bottom-up' approach to nanotechnology which is targeted toward the development of molecular-scale information processing systems."

Later Stoddart converted the shuttle into a "train." He bent the straight track into a squarish, closed loop of atoms, and placed a station on each of the four sides of the square. He positioned two molecular beads (the "trains") on the same "circle line" track, and got them to race each other around the loop. Making three hundred stops per second, each train stopped at all four stations in sequence.

"It is possible to create an apparently complex molecular structure with remarkable ease," he said — and all of it with the techniques of traditional solution chemistry. What new wonders could be expected when mechanochemistry arrived on the scene?

Eric's goal had always been to design not only "molecular machines" but a molecular manufacturing *system.* That was his end-in-view — the *use,* the *purpose* — of all his considerable thought and planning. But you had to build the parts before you could build the whole, which is why he'd started out with the gears and bearings. Still there was this difficulty: How could you be sure that a given molecular design, whether for a gear, bearing, or anything else, would actually and in fact work as intended? There was nothing more common in the engineering world than good-looking designs that failed pathetically when put to the test.

In ordinary macroscale engineering such embarrassments were avoided, to the extent possible, by building a prototype of some sort: a demo, a test model, maybe even a scale model. That was where reality entered the picture, setting up a feedback loop by which the initial design could be modified and improved.

But none of that was possible in Drexler's case because he was engaged not in ordinary engineering but in "exploratory" engineering, which by definition ran far ahead of implementation or actual experiment. There was no chance of testing a given "exploratory" design because such designs were not yet producible and wouldn't be until the tools arrived with which to produce them. The whole enterprise was highly theoretical and removed from practice, and was therefore somewhat at a disadvantage.

"One cannot build systems and hence cannot experiment with them," Drexler acknowledged. "This presents a key challenge."

In the late twentieth century, however, that challenge could be met, after a fashion, for the fact was that there now existed a sort of intermediate stage, a third distinct metaphysical level, between abstract design and real-world implementation: there was the computer model, or simulation. Computer modeling was a large part of normal macroscale engineering, and by the early 1990s every major manufacturer, and even lots of minor ones, were heavy into CAD-CAM: computer-aided design and computer-aided manufacturing.

"Boeing builds models of airplanes on computers before they build them in real life," said Ralph Merkle, of Xerox PARC's computational nanotechnology project. "And the reason they do that is very simple. If you model a process on a computer you can go through more designs more quickly, you can find errors more cheaply and more easily. You can try out systems that either are too expensive or too difficult or too time-consuming to analyze by building a real physical model. As a net result of all these things, the overall development time of the airplane is reduced."

The theory was that simulation processes that worked for airplanes, cars, or toasters ought to work equally well — or reasonably well — for Eric Drexler's molecular machine designs, and so while he was working on his Ph.D. thesis Eric would model some of its primary chemical structures on molecular mechanics and molecular modeling programs. He'd sit there late at night in his stuffing-come-loose easy chair in front of his Macintosh, the one held down by bungee straps on the swing-away shelf, and which was connected in turn to four separate, stacked hard disks; with books, journals, papers, and articles stacked ten or twenty deep on the desk nearby and on the circular coffee table opposite (more in the filing cabinets); with the heaviest, thickest, jumbo-scale reference books stacked eight or nine deep on the floor, plus software boxes, manuals, documentation, papers, et cetera, strewn all over (just your basic reference materials); plus various healthy, ailing, and dead computers, monitors, disk drives, and keyboards scattered here and there; plus cardboard boxes; he'd sit in these tidy surroundings and he'd watch his molecular machines *actually working* — bearings turning, gears meshing — just as if they were right there in front of him, just as though they actually existed.

Which, at least in some tenuous, secondary, limbolike sense, they did. A working simulation bestowed upon his abstract designs a modicum of truth and being, a kind of semi-existence. After all, if such structures appeared on the screen and worked in molecular mechanics simulations, then they could hardly be dismissed as mere make-believe, or as figments of the imagination, or as "science fiction."

Of course, not all of these designs actually worked, even in simulations.

"Basically it's not the case that you put these things together and they always work," said Ralph Merkle. "Sometimes you put them together and model them on the computer and they go *boom!* — right in your face. And you say, 'Oops!' If you want to see a disaster, we've been working on a planetary gear deal that blew up."

But after the usual amount of poking, prodding, and judicious retrofitting of materials and arrangements, the device could usually be made to work, as was finally true even of the planetary gear.

Merkle collaborated with Eric on a number of basic molecular structures. A Stanford Ph.D. in electrical engineering, Merkle had read *Engines of Creation* as soon as it came out, saw Eric give a lecture at Stanford or Berkeley or someplace, and by that time was absolutely convinced that nanotechnology was the next new sensation. Soon he'd persuaded his boss, John Seeley Brown, director of Xerox PARC, to assign him to nanotechnology work on a full-time basis.

"Basically PARC has an approach to its research which is very broad," said Merkle. "There is a great deal of interest in things that have a significant impact, things that would have a potential impact that is large. Certainly nanotechnology has a potential impact which is large.

"Also you have to understand," he added, "that for a corporation the size of Xerox to have one person looking into this, it's an insurance policy."

Of course, simulating molecular machine components was not the same as building them. Many working chemists, indeed, took a rather dim view of the whole notion of molecular simulation.

"What can be done with molecular modeling programs need

not have any equivalent in the real world," said Leo Paquette, an experimental chemist. "Thus the fascination of certain chemists (nonexperimentalists) with the computer!"

"Chemistry is an experimental science, and so full of surprises," said Roald Hoffman, of Cornell. "It is not necessarily true that what the modeling programs give will carry over to the real world. You can build models, but in my opinion it's a long way until we can really trust those modeling programs."

"Everybody has these fancy molecular modeling programs, but they're fairly useless," said Rick Danheiser. "The predictions they make usually are wrong or very unreliable; it's the kind of thing you have to take with a grain of salt. My own experience, and that of other colleagues who use the same sorts of methods to make very simple predictions about the interaction or behavior of very, very simple molecules, is that these things are so lacking in precision and reliability that they're not going to become more accurate and reliable when you're extending it to these huge things."

That was one side of the story. The other side of it was that, despite their limitations, these same molecular modeling programs had been used in a variety of real-world contexts and had worked out well enough. There was the case of structure-based drug design, for example.

Most of the drugs that had ever been produced in recent history had been discovered by accident or through trial and error. Lately, though, a more rational approach to the process had taken root.

"Our target is not the drug, but its molecular target in the body," said Charles E. Bugg, biochemist and a pioneer of the new approach. "We solve the three-dimensional structure of a substance known to participate in some disorder. Then we build a chemical that precisely fits the target and alters its activity."

The term "precisely fits" was meant literally: in many cases it was the actual physical shape of the drug molecule that determined whether or not it worked. Bugg and his collaborators had worked on a drug that would block a certain chemical reaction that played a role in cancer.

"We first determined the three-dimensional arrangement of

the target's constituent atoms, paying particular attention to the active site," he said. "Next we turned to our computers."

The hope was to simulate various drug candidates on the computer, "testing" them out there before experimenting with them in the lab.

"As we viewed a candidate on a monitor, we worked it into the active site, examining how well the shape and chemical structure of the candidate would complement that of the site. We also used programs to help us estimate the strength of the attractive and repulsive forces."

The drug developers were aware that their modeling programs were not perfect, but they also knew what their shortcomings were, and so it was possible to adjust for them. It was also possible to home in on a drug by means of successive approximation: going back and forth between computer simulation and experimental test in the lab.

"This iterative strategy — including repeated modeling, synthesis, and structural analysis — led us to a handful of highly potent compounds that tested well in whole cells and in animals."

The long and short of it was that a drug-development process which normally took ten years and cost millions of dollars was completed in less than three years and at far less expense. But if computer modeling of chemical structures worked for pharmaceuticals, then why wouldn't it work for nanomachines?

Ralph Merkle was convinced that it would, especially after he'd "discovered," from his modeling work, a little-known fact about nature — the existence of dislocations.

"Dislocations have been studied in material science since the fifties," said Merkle. "Actually it's amusing how I got this particular dislocation. I didn't know the dislocation literature, but I knew I wanted something that would let me bend diamond."

Because it consisted of carbon atoms, which formed tenacious three-dimensional covalent bonds, diamond had long been Drexler's material of choice for nanodevices. It was strong and rigid, the hardest stuff known to man, and so Merkle did a lot of work with it. The trouble with diamond, though, was that its molecules wanted to spread themselves out in flat planes, which of course presented a problem if what you wanted to do with it was to bend it into a ring, tube, or other curved structure. You could force it to do such a thing, but only at the price of subjecting it to some rather unsavory internal stresses and strains.

The solution was to add in some extra bits of matter — extra atoms — around the outer surface of the tube, much as a bricklayer added in some extra cement and chinking along the outer surface of an arch. Those out-of-order atoms would relieve the strain, allowing the flat surface to assume a curved shape naturally. In materials science, such out-of-order atoms went under the heading of "dislocations."

"So I sat around and played with it over a weekend, until I got a dislocation that I liked," Merkle recalled. "Then on Monday morning I went down to our materials science group here at PARC and brought my dislocation up on the screen. I said, 'I like that dislocation. What is it?' And they said, 'Oh, that's a Lomer dislocation, discovered by Mr. Lomer or Dr. Lomer back in the 1950s.'

"And then I tracked down the references on it. When you talk about dislocations, people often think of them as being very active, having dangling bonds, being horrible and awful. But the Lomer dislocation is well studied, and it's a stable dislocation. There are no dangling bonds, every carbon atom in the structure has four neighbors bonded to it, it's stable, it's reasonable, everything is hunky-dory."

And if it worked on the screen, why not in the machine?

Building an actual molecular machine was something that Drexler himself, the dedicated theorist, was not exactly suited for. "I wouldn't know which end of an STM is up," he often said. There was a division of labor in science just like anywhere else, and so while he took it upon himself to design the machines, he left it to others to actually make the things.

Of course, he tried to help matters along somewhat. Starting in the fall of 1989, Chris and Eric held "Foresight Conferences on Nanotechnology," the purpose of which was to bring the right people together — the experimentalists, the actual hands-on workers, the technology developers — in order to get the revolution started.

Handpicked and personally invited by Drexler himself, some of the experts had never even heard of nanotechnology, much less Eric Drexler, and could occasionally be heard asking each other questions like, "So what's 'an assembler'?"

By the time of the second conference, two years after the first, the essential path-breaking nano-events were securely in place: Bill DeGrado's engineered protein; IBM's atomic logo and the one-atom switch; Rebek's self-replicating molecules; Stoddart's shuttles and trains. As one speaker after another made clear, some further gains had been made in the interim.

Kumar Wickramasinghe came to the conference from the IBM Thomas J. Watson Research Center and told one and all that STMs had detected not only atoms but also individual electrons and even the impalpable van der Waals forces.

"You can use STMs to generate sound by tapping the tip on a surface," he said. "You can detect forces smaller than Brownian motions, at 10^{-3} angstroms. This is pretty significant work."

David Blair, from the University of Utah, told about the atomic-scale motors that already existed in nature, in *E. coli* bacteria, whose propulsion system consisted of flagella, twisting screw-like devices turned by motors that were so tiny they just barely showed up in electron microscopes.

"The motor is driven by 'protonmotive force' using 1,000 protons per revolution," he said. "RPM is 6,000 at room temperature; at 37° Centigrade it cranks at 18,000 RPM — that's redline. Power output is one-tenth of a micro, micro, *micro* horsepower, 10^{-19} horsepower. It turns out that if you had a mole of flagellar motors — if you had an Avogadro's number of them, 6×10^{23} — that would be about right to power the Queen Mary."

A working electric motor less than half the size of Drexler's assembler — that was a proof of concept for you.

Anyway, after three jam-packed days of this stuff, Eric sent his small army of new friends and associates back out into the world: "My hope is that when this meeting closes, it will be followed by an open-ended series of informal and private meetings, which will be held among researchers in collaborative efforts to break new ground, both in computational modeling of systems and most especially in the laboratory, in developing the capabilities and devices that are necessary to open this new realm of science and technology."

The conference was written up by the *New York Times*: "Atom by Atom, Scientists Build 'Invisible' Machines of the Future," the headline claimed. The story quoted Drexler as saying, "We've built up this vast storehouse of basic capabilities. We are on the threshold of molecular manipulation. This is close enough that I can taste it."

Unfortunately, just when he'd imagined that the whole nanotechnology enterprise had gained a new respectability, that the "cultishness and bullshit" of yore had been banished from the scene, there seemed to be a fresh new outbreak. One afternoon, as people filed out of the conference center, each of them was handed a flier that said at the top:

ARE MOLECULES SACRED?

Eric took a look at that and rolled his eyes. Here was this bunch of hard-science types — earnest workers, laboratory people — and now they were one and all being asked to contemplate this metaphysical, semi-religious mystical conundrum: "Are Molecules Sacred? If not, what is? — if anything? What is worthy of deep respect and honor?"

Written by BC Crandall, who'd later coedit the conference proceedings for MIT Press, the document was apparently the world's first manifesto for a *truly* deep ecology, one that reached all the way down to the innermost constituents of matter.

But why not? Once you'd realized, as the nanophiles had, that there was a complete new world down there — virgin territory, wholly uncharted and underutilized realms — well, the question of *exploitation* arose, didn't it? Was there any good reason not to treat those wee atomic creatures with the same rights and respect, the same care and concern, with which the enlightened among us now regarded animals and trees, the Chesapeake Bay and the Grand Canyon?

Didn't molecules have rights, too? Were we going to discriminate against them just because they were *small?*

"Perhaps molecules deserve our reverence and deep gratitude," the handout said.

Farsighted as he was, Drexler had never imagined it would come to *this!*

Eric got no flak from the Media Lab. Minsky would be chairman of an interdepartmental committee that would supervise Drexler's research, as well as give him written and oral examinations in physics, chemistry, and computer science. In addition to Minsky, the committee would consist of Rick Danheiser, chemistry; Alexander Rich, biology; Jerry Sussman, EECS; and Steven Kim, mechanical engineering.

Until the day it was over and done with and Eric walked away with his prize, all of this would be kept a big secret. No one in the outside world, and few even within MIT, were to know.

"I had been publishing scientific papers, and writing books, and giving talks, and talking to corporate research laboratories, and getting money for allegedly being an expert in this field for many years," Eric explained later. "To have the visible status of graduate student while doing all this seemed to me to be a bad idea. And to have the visible status of graduate student when the whole program might not work out for one reason or another — look what happened in EECS — seemed to be an even worse idea. So I said, well, there's no particular reason to have this talked about until it's done."

In fact, it was not talked about in some quarters of MIT even well after it was done. Six months after the degree had been awarded, Robert Silbey, head of the MIT chem department, leaned over toward a visitor and said in hushed tones: "You know, I'm not supposed to say this . . . but Eric Drexler is getting his degree here."

Eric's thesis, "Molecular Machinery and Manufacturing with Applications to Computation," was unique at MIT . . . or anywhere else, for that matter. It wasn't the usual "salami science" thesis in which you took some tiny field and made the thinnest possible new advance; instead it laid out the defining principles of a new technology, a new order of things.

Not that this made any difference to the examining committee, few of whose members would admit to having read it.

"Oh, I . . . put it this way, I read it the way I read all theses that come to me," said Alexander Rich, the biologist. "It didn't have much biology in it."

"I have not read the whole thing," said Rick Danheiser, the chemist. "I have perused all of it — I have at least skimmed all of it — and some parts I have read in detail."

There were various opinions as to its quality.

"It stood on its own feet as a reasonable dissertation," said Rich.

"It was an extremely nice thesis that turned into an excellent book," said Sussman.

"It's very impressive because he talks about a number of things which are familiar to me and describes them very succinctly, and from a perspective that I find quite interesting," said Danheiser. "Even though it's a field close to my own, I couldn't have done a better job. And in addition he talks about things I know nothing about at all. So it's very impressive, there is no question."

"The thing that bothered me about it was that it showed utter contempt for chemistry," said Julius Rebek — he of the self-replicating molecules. Rebek was not on Drexler's committee, but Danheiser gave him a copy of Eric's thesis anyway. He thought he'd like it.

Fond hope!

"The molecules that he drew as these wheels and so on were things that are far beyond bonding rules as we know them, and beyond the ability of anybody to even think about making them with today's techniques," said Rebek. "And the mechanosynthesis stuff that I saw in that thesis might as well have been written by somebody on controlled substances." (Julius Rebek, a Hungarian, has been known to exaggerate for comic effect.)

"Well, but Rebek didn't *read* it!" said Danheiser.

"I read *part* of it!" said Rebek.

Eric's thesis defense was open to the public, as was normal practice at MIT. Absolutely anyone in the world could attend, whether they were connected to MIT or not, and could ask the candidate any and all questions they wanted to. On the day of the defense, which was held Friday, August 9, 1991, in E15-054, a classroom in the lower level of the Media Lab ("the building that looks like an elephant," said Rebek), the committee members, except for Alexander Rich, who was out sick, showed up, together with some of Eric's friends and supporters from the NSG — "these guys trapped in the sixties or something," in Rebek's view. "You know, with sort of a glazed-over look."

Eric's critics from EECS, unfortunately, were not on hand for the final showdown. Julius Rebek was, although he'd rather have been in Philadelphia. "Rick dragged me along," he explained. "He talked me into it."

Drexler gave a formal presentation of results, after which he entertained questions.

"A presentation is usually quite good," said Gerald Sussman. "In Mr. Drexler's case it was *extremely* good."

As for the defense itself, in classic *Rashomon* fashion everyone had their own separate perceptions as to how it all went.

"I was the only one there who ever asked any hard questions," said Rebek. "And I could tell that by and large people didn't appreciate me asking hard questions." (Eric's fans were on hand, after all.) As for Eric's answers to his hard questions, "He blabbered on about whatever," said Rebek, in his fashion.

"There *weren't* any serious objections," said Minsky. "There was some discussion about the rigidity of certain chemical bonds, but it wasn't a serious argument because if that particular structure wasn't rigid there were lots of others you could use."

"It was a standard examination," said Sussman. "It was no more or less than average. I think it was quite similar to others that I have participated in during my twenty years, almost, of being a faculty member."

After the normal two hours, the committee held a brief private meeting and approved both the thesis and the defense, and five weeks later, on September 18, 1991, the Massachusetts Institute of Technology formally and officially conferred upon Kim Eric Drexler the degree of Doctor of Philosophy, "in the field of molecular nanotechnology" — as the diploma stated in black and white.

A rather exotic item, that diploma. It was the first degree ever awarded in the subject, anywhere in the world. And it was all the stranger because MIT had no school, division, or department of "molecular nanotechnology," nor had it ever offered any courses in the subject. Drexler, moreover, had been enrolled the whole time in MIT's "Course 4," otherwise known as the Department of Architecture, part of the School of Architecture and Planning, the administrative home of the Media Lab. Truly, it was an organizational mess of grand proportions.

Not that the Institute's higher-ups couldn't put the best face on things.

"One of MIT's great strengths," said Frank Perkins, dean of the graduate school, "is its ability to tolerate such ambiguity."

14 The Greek Chorus of Woe

Nanosystems: Molecular Machinery, Manufacturing, and Computation, came out in the fall of 1992, published by John Wiley and Sons, of New York. This was Eric's technical text, his fully-fleshed-out blueprint for nanotechnology, and also his answer to *Limits to Growth.*

Toward the end of *Nanosystems*, in chapter 14, after some four hundred pages of chemical, physical, and computational analysis, and after roughly an equal number of equations, plus charts, graphs, diagrams, and the like, the reader came upon a section entitled "An exemplar manufacturing-system architecture." As in the rest of the book, the discussion here was so abstract and technical, so carefully couched in numerical estimates, and so dependent upon the even more complicated stuff that preceded it, that many readers doubtless never got the full force and import of what it was they were confronted with when finally, toward the end, on page 421, they came to the section-heading "An exemplar manufacturing-system architecture," and the first paragraph beneath it, which read:

> This section and Table 14.1 outline a system capable of manufacturing macroscopic objects. The subsystem capacities are chosen to permit the conversion of a feedstock solution consisting of small organic molecules into ~1 kg product ob-

> jects of ~0.2 m dimensions in a cycle time of ~1 hour. The flow of materials proceeds through molecule sorting and orientation (Section 13.2), preparation of reagent moieties (Section 13.3), several stages of convergent assembly using mill-style mechanisms (Sections 13.3 and 14.2), and several stages of convergent assembly using manipulator-style mechanisms (Sections 13.4 and 14.2).

It was the most arcane and boring paragraph imaginable.

Some clues as to what it actually meant appeared later on, where it said that this exemplar manufacturing system could "make almost any macroscopic shape in ordinary use today within better tolerances than those provided by conventional manufacturing," objects "that exhibit material properties that equal or better those of present industrial products."

The astute reader also gathered that the exemplar manufacturing system would have a fan for cooling — but because this was Eric's technical text, written for scientists, engineers, and technology developers, he could not simply *say* "a fan for cooling." Rather he said: "The required cooling capacity can be provided by fan-driven air flowing at a rate of 0.1 m^3/s with $\Delta T = <15K$ between the intake and exhaust ports."

So it took some careful reading, and some mental translation into everyday English, to get the full and wondrous message. But when you did, you found out that what this exemplar manufacturing system was really all about, what it actually did, was to take an incoming stream of raw materials and, in less than an hour, transform them into real, physical, solid objects. Normal objects that you could use in real life. This, finally, was Drexler's molecular manufacturing machine, the in-home factory, the little black box that would make for you, atom by atom, everything you ever wanted.

When he described it to you in conversation, you immediately got the picture: "This is a box about the size of a microwave oven, with four rubber feet and a fan to keep it cool, that plugs into the wall, that would have a tank on the top into which you would put feedstock solution, a special mix of chemical compounds like acetone — the material in nail-polish remover — as a carbon source, and some other things dissolved in it to supply other elements. You'd be able to pick some object from a catalog and say, 'I want one of these,' and have it work for an hour or so. And then it would give you about a kilogram of product.

"It's an appliance which if you turn it on and give it the raw materials and press the right buttons, will produce any object in the catalog" — as for example, a television, a dishwasher, or an atomically perfect copy of itself.

"The range of products that can be produced is large," Drexler wrote, "encompassing high-performance structures, massively parallel supercomputers, and additional molecular manufacturing systems."

This was the desktop manufacturing unit, the universal constructor, the molecular cornucopia. It took no talent to program it, nor human labor of any kind: "Between input of raw materials and the removal of finished products, no labor is necessary."

Nor, in most cases, was an external power source. The unit plugged into the wall not to *draw* electrical power, but to *get rid* of it, to carry off the excess energy liberated by the chemical transformations taking place inside: "This surplus energy could be conveniently disposed of by delivery to an external power distribution system in the form of electrical energy."

Everyone alive could have one of these little black boxes, the reason being that the units could churn out copies of themselves as easily as they could churn out anything else. Computers to cashew nuts — it was all the same to the machine. The world could look forward to an era of sustainable growth, responsible use of natural resources, stunning reductions of pollution and waste — and all from these little desktop fixtures. Each machine would use, as raw materials, only as many individual atoms as were required to make the desired object. It was the ultimate clean machine.

Not a word of it ever got out to the press.

Nothing appeared in *Time*, *Newsweek*, or anywhere else announcing that there was now in existence a reasonably well worked-out technical plan for an item that could provide everyone living with virtually all of their necessary (and unnecessary) material goods, at negligible cost, and with little if any expenditure of human labor or electrical power.

The molecular cornucopia was on the drawing boards, the plans were there for all to see, but no one was informed of the fact. Apparently, either the thing was impossible, or it was unbelievable, or it was just not news.

Critical evaluation of nanotechnology was divided along lines of whether it was pre- or post-*Nanosystems.* Those scientists who felt they knew enough about Drexler to venture a pre-*Nanosystems* opinion tended to dismiss the whole idea as mere hand-waving, as idle boasting, as the science-fictional fantasies of an overactive imagination.

"There was this movie a quarter-century ago with Raquel Welch and, you know, submarines going through your bloodstream and doing things, and this is a somewhat dressed-up version of that stuff," said the Anonymous Critic, a Harvard Ph.D. "I'm not willing to criticize — and now I'm characterizing Drexler — material which is basically science fiction. There's no discipline in it whatsoever. I'm not willing to criticize it as there'd be no purpose in it. It'll fall of its own weight, eventually. I think it's all hype."

Such a viewpoint, which was entirely typical, tended not to be based on what could be called a searching personal study of Drexler's various works. The Anonymous Critic, for example, hadn't read the technical papers or even *Engines of Creation.*

"I believe I've had it in my hands," he said. "I probably glanced at it. Have I looked at every word in it? Undoubtedly not."

What, then, was he going by? One of Eric's lectures, primarily.

Often enough it seemed that Eric made as many enemies as friends when he traveled around the country giving talks before technical audiences. Or at least that was true pre-*Nanosystems.* There was the time he went up to the IBM Thomas J. Watson Research Center in Yorktown Heights, New York, for example, to tell them about nanocomputers and rod logic.

"I got some some relatively sharp questions," Drexler recalled later. "My sense was that the questions had reasonable answers. There was one specific question about something involving transformations of carbon — chemical transformations with carbon and mechanical force — that was formulated in a way such that I think the answer that I gave might not have been terribly clear or wonderful. That's the main concrete negative thing that I can recall. But on the whole my reaction was that here were a bunch of chemists and physicists who are listening to a presentation of ideas, they throw darts at them and the darts don't stick. And my default assumption in those circumstances is that people go away saying, 'Gee, maybe there's something to this. There seem to be answers to the questions. Maybe it stands up to criticism more generally.'

"I gather that's not a universal response, however."

Some in the audience, indeed, remembered the event quite differently.

"His lecture was received very poorly," recalled Shalom Wind, who attended it. "People were laughing in the hallways afterward."

Shalom Wind, one would think, would be a big fan of Eric Drexler — or at least of nanotechnology. His IBM business card bore the legend "Nanostructures and Exploratory Devices." He had a Ph.D. in physics from Yale. He was doing actual, hands-on lab work at the nano level. But he didn't think much of Eric Drexler.

"He hasn't invented anything," said Wind. "He hasn't demonstrated it. Has he patented any of this? No. If Drexler can demonstrate the germ of his ideas, then he would be better accepted."

"To me, it sounds like it's impossible," said George Sai-Halasz, another IBM Watson researcher who'd attended the talk. "I mean, there's nothing on the horizon that would point in the direction that this is going to work."

Besides which, Drexler's stuff was so sketchy, and written so much on a popular level, most of it, that it was hard to regard it as serious science.

"He apparently does not have a lot of papers that have been reviewed by his peers, or by people who are accepted experts in the field," said Shalom Wind. "This is America, so anybody can do what he wants, but a guy who has published two books already, before he's had the kind of critical review of experts in the field, his audience has clearly been the public, not the scientists."

And this charge, too, was not without foundation. Drexler had published a small number of technical papers, most of them in out-of-the-way journals, but at least pre-*Nanosystems* the bulk of his stuff was on the popular level. After he'd published *Engines of Creation*, moved to California, set up the Foresight Institute, and gotten his visiting-scholar position at Stanford, the logical next thing for Drexler to do was to finish and publish his technical text. That being the logical next step, obviously the right thing to do, Drexler again did something else. He helped write a *second* popular account of nanotechnology, *Unbounding the Future.* Written largely by Gayle Pergamit, a science writer, and Chris Peterson, his wife, Eric worked on it too, but his name came first on the cover — "for marketing reasons." The book came out in the fall of 1991.

Even among Drexler's staunchest supporters, nanofans, and friends, *Unbounding the Future* raised a few eyebrows. It broke no new ground, was less challenging and interesting than *Engines,* and was almost in every other way a disappointment. Reaction outside the nanoworld was even less approving. "Not only didn't it add anything to the field, it even subtracted something," said an editor at the *New York Times Book Review,* explaining their decision not to review the book.

Why then did they write it? "People told us that *Engines* was too technical," Chris Peterson said. "We wanted to write a more popular account, one that people could understand more easily."

Unbounding the Future, though, was a popularization of a popularization — not the type of thing calculated to make the best impression on working scientists. It was even worse that for some unknown reason many of these same scientists had gotten it into their heads that Drexler was out there raking in grant money by the ton — government money, foundation money, *their money!* — and all of it on the basis of these fleeting science-fictional visions of his.

"Drexler's able to sell himself very well," said Vince Rotello, the MIT chemist. "He's able to get money. Drexler's able to get money for things that I don't think will ever work."

"Once somebody wants to direct resources in a certain direction, then certain things have to be justified," said IBM's George Sai-Halasz.

"If our government is funding the kind of work that he's doing, and if we submit a proposal that ends up not being funded because *his* work is being funded, I would severely, or strongly, object to that," said IBM's Shalom Wind.

"Here at IBM, this is where people will start to get upset, when you start talking about money," said John Foster of the IBM Almaden Research Center.

Was that the rub — grant envy? His is bigger than mine? Visions of K. Eric Drexler, the Major Corporate Giant, grabbing large amounts of federal cash out of the hands of poor little dwarfish IBM?

Well, it was hard to shed any tears over that. Drexler would later claim that he'd never even submitted a proposal for federal money, much less gotten any, but that was not strictly true: he'd won a National Science Foundation fellowship as a grad student. But that was all he ever got.

Still, money wasn't really the problem.

"The problem that people doing today's research have with Eric Drexler is that he's looking so far in the future that it's kind of out of sync with what they're working on presently," said Tom Newman, the man who'd won the second Feynman prize. "So you'll find some people who are irritated by Eric Drexler. And I think that's true of some people maybe even in this building."

"This building" was the IBM Thomas J. Watson Research Center, perched high up on a bluff, an overlook, where Newman now worked. Many of those who worked there, indeed, were extremely protective of the place, as if the slightest blip on the horizon would send the whole structure crashing down to the duck pond below, into which it would disappear from the face of the earth, forevermore.

"We're very proud of the way we do our science," said Shalom Wind. "IBM Research has a reputation among scientists around the world for doing some of the finest things that have been implemented in technology. The dynamic RAM cell was invented here."

"The electron beam as a tool for writing very small structures basically came out of this building," said Rolf Landauer, himself an IBM Watson trailblazer. "A lot of the interesting physics that you get when you make structures very small have come out of this building. The use of periodic structures — resident tunneling, quantum-mechanically building periodic structures with molecular deposition, the two-dimensional electron gas, a lot of the physics of playing with electrons, small structures — all that's come out of the work in this building. And we've known that, we've worked hard at it. The scanning tunneling microscope has come out of this organization, IBM."

So who was this *Eric Drexler* character, this guy with his pop science and pipe dreams, who was *he* to be telling the people *in this building* what the future of computation was? And you could understand their point perfectly well.

Anyway, with the exception of the *Whole Earth Review*'s "Critique of Nanotechnology," it was difficult, pre-*Nanosystems*, to find anyone in mainstream science who took nanotechnology seriously enough to give it anything more than the idle brush-off.

"I suppose it would be a little hard to find critics right now because in a sense it's still just a little too close to the pie-in-the-sky stage," said A. K. Dewdney, the *Scientific American* columnist. "Who can criticize something this early?"

During the pre-*Nanosystems* era, in fact, the single published criticism of Drexler to appear in a technical journal was a couple of sentences in an article by Kurt Mislow, the Princeton University chemist. Supposedly, according to rumor, this was a Big and Telling Criticism, one that, if true, would undercut Drexler's program in a major and fundamental way.

The article in question, "Molecular Machinery in Organic Chemistry," was published in the specialized journal *Chemtracts* in 1989. Mislow, the author, was the world's top expert on molecular gear systems, examples of which he'd synthesized using the methods of traditional solution chemistry. These gears, he said, "resemble to an astonishing degree the coupled rotations of macroscopic mechanical gears." They also resembled little propellers — fan blades, tiny paddles — but they were actually molecular gears, he said, examples of what Drexler had been talking about for more than a decade.

"It is possible to imagine a role for these and similar mechanical devices," said Mislow, "molecules with tiny gears, motors, levers, etc., in the 'nanotechnology' of the future."

Nevertheless, Mislow, who'd actually read *Engines of Creation,* saw a problem with Drexler's account of molecular bearings. Drexler imagined that molecular bearings would "turn freely." This, said Mislow, was not possible.

"Uniform angular velocity implies absence of energy barriers," he wrote in *Chemtracts.* "This state is unattainable for rotors in chemical systems: There is no such thing as 'free rotation'!"

In answer to which, Drexler denied that he'd ever said there *would* be such a thing as "free rotation." Although in *Engines of Creation* he'd said, "Conveniently enough, some bonds between atoms make fine bearings; a part can be mounted by a single chemical bond that will let it turn freely and smoothly," he was not speaking of free rotation in the technical sense used by chemists.

"Saying that a bearing, a wheel, a doorknob, or anything else can 'turn freely' has a simpler, plain-English interpretation," Drexler explained. "It means that the motion isn't blocked or subject to excessive friction. Good nanomechanical bearings can turn freely, with remarkably low friction. 'Free rotation' was never claimed."

Which, in response, Mislow seemed more than willing to accept: "Drexler may have used the phrase 'turn freely' as a figure of speech and may not have meant it literally. In Drexler's article 'Nanomachinery: Atomically Precise Gears and Bearings,' he does not talk about free rotation; instead, he shows his awareness of atomic 'bumpiness' and suggests ways of minimizing this kind of friction."

And that was the end of the Big and Telling Criticism.

The problem pre-*Nanosystems,* then, was not that there was too much criticism of Drexler. The problem was that, other than for Kurt Mislow and the *Whole Earth Review,* there was no criticism whatsoever.

"It seems to be no mystery in our office," said an editor at the British science journal *Nature.* "My colleagues tell me that there are perfectly good reasons why he is disparaged by the rest of the community. For example, he is supposed to have an almost universal disregard for the chemical and physical properties of the things he proposes to manipulate."

Well, that's what you got for writing popular stuff.

Any lingering impression that Drexler had "an almost universal disregard for the chemical and physical properties of the things he proposes to manipulate" was instantly dispelled by *Nanosystems,* with its five hundred pages of dense technical prose, most of it having to do with such timely matters as "continuum models of van der Waals attraction," "interfacial phonon-phonon scattering," "bearing stiffness in the transverse-continuum approximation," "alkene and alkyne cycloadditions," "finite-state machine structure and kinematics," and the like. The text itself was aided and abetted by innumerable graphs, charts, diagrams, and tables, logarithmic plots of "root mean square displacements of harmonic oscillators of varying mass and spring constant as a function of temperature," "dimensionless transverse variance for rods, neglecting shear compliance," and so on. There were figures showing items of such major interest as "a series of structures modeling a diamond (111) surface, with corresponding values of k_{sz} (MM2/CSC) for vertical displacement of the central-surface carbon atoms with respect to the lattice-terminating hydrogens below (shown in ruled shading)."

There were huge, overpowering mountains of the stuff, avalanches of facts, absolutely an inconceivable volume of sheer physical and chemical information.

And then, finally, you got to the meat of the book. Which was to say, the equations.

They were something else, those equations. There were no less than 450 separate equations, formulas, and various collateral inequalities, variations, and transformations scattered throughout the text.

It was as if he'd wanted to rub people's noses in them. Whenever he gave lectures before scientific audiences Drexler made a point of including a slide or two of the most imposing relevant equations he could find. He'd get up there on the stage and talk for a while and then he'd project this slide on the screen — it would have three or four extremely complicated formulas on it — and in his hollow, flat, fluting voice he'd say something to the effect, "Here are a few of the parameters involved," and he'd actually even explain them for anyone who was listening, but what he was really doing with this, what he was *really* saying, was: "Look, folks! I'm not just dreaming this up! I do know a bit of the math involved!"

And so it was with *Nanosystems.*

By any standard it was an extraordinary document. The Association of American Publishers named *Nanosystems* the outstanding book in computer science for 1992. In May of the following year, Drexler won the Kilby Young Innovator Award for 1993. Chris and Eric flew to Dallas for the grand reception — Eric even bought a tuxedo for the occasion, spending actual folding cash on clothes! — there to meet Jack Kilby, coinventor of the integrated circuit, after whom the award was named.

Even Drexler's critics — some of them — allowed as how *Nanosystems* had maybe changed their minds . . . just slightly.

"There's more to this than I thought," said the Anonymous Critic. (This same critic had previously sent back a prepublication copy of *Nanosystems,* explaining that, "as an industrial employee I cannot afford to look at material which could, conceivably, have proprietary content. I know it is, very likely, all innocuous. On the other hand it is quite possible that just about now some employee who does not know me and has never heard of Eric Drexler is submitting an invention which . . . Therefore, I am returning the material uninspected.")

"There's some interesting stuff there," said Shalom Wind, about *Nanosystems.* "He's done an awful lot of work, I have to admit."

"Parts of it are very engaging reading," said Harvard chemist George Whitesides.

"Drexler is remarkably well informed on a broad range of subjects," said Kurt Mislow. "He is amazingly au courant with the latest developments in organic chemistry and, so far as I can judge, molecular biology. His treatment of these subjects is thorough, accurate, analytical, and sophisticated."

And there were even some people at IBM Watson who, not wanting to be thought gauche, or to speak irresponsibly or recklessly, or to live dangerously, or to embarrass themselves or IBM or the consecrated memory of Thomas J. Watson — or, God forbid, to *lose any federal money!* — they'd say to you, over the telephone: "You don't have your tape recorder on do you? . . . No? Well, good, because I don't want to have my name associated with this, but . . . you know, I think that even as far-out as Drexler's stuff is, I have the feeling that when we finally get down to do actual engineering at these regimes, that the calculations he's done . . . they may turn out to be right."

Well! What a loose cannon up there! Quite unrestrained! Quite a wild outburst from the ranks of IBM Watson!

There were criticisms, however, especially from chemists, some of whom didn't like the style of the book or any of its proposed structures. Others claimed that those structures were too complicated . . . or too simple . . . or that they were chemically unstable . . . or that even if they *were* stable, they would never work as advertised . . . or that even if they *did* work as advertised there was still no way of actually *making* the goddamn things . . . *So why all the bother*, for crying out loud?

"The book is fussily subdivided into section, subsections, subsections of subsections, and so forth, each providing the reader with an overwhelming wealth of details," said Kurt Mislow. (This was a criticism.) "I felt myself drowning in a sea of factoids and in a torrent of words. Drexler seems to be of the party that believes: Why say in 10 words what you can say in 1000?"

Mislow was particularly offended by a paragraph in which Drexler listed an admittedly rather breathtaking collection of molecular devices that could be constructed by his tiny assemblers: "It is feasible to construct nanomechanical rotary bearings, sliding shafts, drive shafts, screws and nuts, power screws, snaps, brakes, dampers, worm gears, constant-force springs, roller bearings, levers, cams, toggles, cranks, clamps, hinges, harmonic drives, bevel gears, spur gears, planetary gears, detents, ratchets, escapements, indexing mechanisms, chains and sprockets, differential transmissions, Clemens couplings, flywheels, clutches, Stewart platforms, robotic positioning mechanisms and suitably adapted working models of the Jacquard loom, Babbage's Difference and Analytical engines, and so forth."

"That's what I like best: *'and so forth,'* " said Mislow. "You know, in case he left anything out, hadn't thought of it: *'and so forth.'* This is ludicrous. It's ridiculous. You can't make these claims without some sort of justification. What is the justification for every single one of those things?"

There were blanket denunciations of Eric's competence in chemistry.

"*Nanosystems* was a big disappointment for me!" said Fraser Stoddart — he of the molecular shuttles and trains. "In it, he reveals that his knowledge of chemistry is scant to say the least. I am sure he would have avoided many of the bizarre suggestions in the Wiley book if he had circulated the manuscript around the community of supramolecular chemists and nanochemists. I welcome and support his vision but I find his mission, when it comes down to the nuts and bolts of the business, lacking in substance and credibility."

"The route to 'nanotechnology' proposed by Drexler is unnecessarily bound up with preconceptions carried over from the macroscopic world," said Vince Rotello. "At best his proposed structures are overly complex, at worst, untenable. The use of more, if you'll pardon the term, *organic* structures would greatly facilitate the cause of nanotechnology."

"I personally do not feel that the designs that Eric has been generating are particularly interesting or significant," said Rick Danheiser. "At this time our fundamental understanding of molecular interactions does not permit the design of specific structures of this sort."

"The molecular machinery suggested is dull," said chemist Roald Hoffman, of Cornell. "It's the same as bigger structures, just tiny. I would look for new features, something not done by classical gears."

"Drexler's perspective, far from being prophetic, is actually retrogressive," said Kurt Mislow. "These machines of his are all mechanical devices: they're crude, almost medieval, I would have to say — hopelessly medieval machinery.

"It seems to me if you wanted to actually make a revolution in this field, you would not only think in terms of reducing the macro to the nano but you would actually *think nano.* If you think nano, you are dealing with a whole different kind of world. You don't deal with little teensy gears and little teensy motors and stuff like that. The crudity of these machines I find just silly, this literal transposition of our macro machines to the microworld."

Julius Rebek saw problems with Drexler's use of carbenes, highly reactive, carbon-containing molecules. So reactive were carbene molecules, he said, that a structure composed of them would reach back around and bond with itself, with its very own molecules, a phenomenon known to chemists as "reforming."

"There is precedent for carbenes," Rebek said. "They're stable when you have them really cold and insulated so that there's one molecule here and it doesn't see another molecule for miles. And its internal structure is such that if you keep it cold enough it doesn't bite back and shrug itself: it doesn't have enough energy to do that, it's like frozen in time. So yes, there is such an entity.

"But structures that he was proposing were *sheets* of carbenes — like next to one another — when such structures are unknown. Anything of that unhappiness reforms its surface. It just shrugs itself into something else."

Eric's doughnut-shaped structures — the "overlap repulsion bearings," the "strained-shell sleeve bearings," and whatnot — these were regarded by chemists with a special brand of molecular horror. Each critic had his own private name for them: "the big macrocycle-type thing," "the Dunlop tire," and so on. Red lights flashed in their eyes, fireworks exploded, all kinds of brain seizures took place in their heads whenever chemists looked at the things. They wanted nothing whatsoever to do with them! They *hated* them!

"The Michelin tire he has with all the nitrogens pointing the same way," said Rebek, ". . . the thing is that all of these electrons repel one another, they want to get away, and when you line them up all one-way you're asking for serious trouble. In fact, molecules like hydrogen peroxide, which have electrons on neighboring atoms, are really explosive, just because of that: *I want to get out of here!* So when you string fifty of these together, you know, good luck."

"He might have a ring of twelve nitrogens linked one to another, whereas so far when chemists have tried to make rings or chains of nitrogens, far fewer can be made," said Rick Danheiser. "Once you have five or six the thing becomes extremely unstable — explosive and so forth.

"What Eric would say, and it's a perfectly appropriate and correct argument, is that if this intrinsically unstable object was insulated from any other matter — by being in a total vacuum and being rigidly held so that no other molecule can come into contact with it, except maybe something very specific, which is itself designed not to interact — then no problem. And that's true.

"Nonetheless, he's designing things which have intrinsic instability. Why do that? At one point I said, 'Eric, it's as if you were proposing that one build an aircraft out of sodium metal — which of course reacts explosively with moisture — by saying that we have this lacquer that we can coat it with that will insulate it so that water will never come in contact with it. So it will be fine, and it's lighter or has this or that desirable property.' I would say: 'Great! Your company can make the airplane out of sodium. But *I'm* going to use aluminum or titanium, because even if you can coat it that way, how are you going to construct the thing? At every step you're going to have to worry about insulating it. And, you know, what if there's an accident?' That's the kind of thing we're dealing with here."

"I could not think of a way of *making* gears," said George Whitesides, the Harvard chemist. "I emphasize the word *make*: I think I could write paper structures for systems that would satisfy many of the criteria for micromechanical systems without too much problem; the issue is to convert them into reality."

"Nobody is talking about making them because there *is* no way to make these things, these various molecular gears and bearings," Rick Danheiser said. "One might just as well develop designs for interstellar spacecraft. The two problems are very similar with regard to our present state of knowledge.

"Someday, I believe, our understanding of physics will allow us to travel to the stars. Someday, I believe, our understanding of chemistry will permit us to create nanotechnological devices and systems. I'm not sure which will happen first."

What a Greek chorus of woe it was! The carping! The moaning! The gnashing of teeth!

But that was the way of the paradigm shift. Those moans were in fact the last gaspings of the outgoing body of thought, a viewpoint that would soon be extinct. Basically it was the conception of the atom as fluff-ball, as the locus of uncertainty, as this wee mystical widget at the bottom of things. Prior to 1990 or so, anyone who was so naive as to oppose the fluff-ball view with a conception of the atom as billiard ball, as a physical object that could be touched, held on to, and mechanically moved — well, such a person was not of sound scientific mind.

Even in the 1980s, atoms were still pretty much theoretical entities: they were real, but real in their own way, meaning that they were wholly unlike Big World objects. Their physical locations were uncertain, and only probable at best; they had no distinct boundaries; they existed only in great statistical heaps and collections, never as clearly defined "objects."

In the face of which doctrines, Drexler's theory had peaked too soon. It was his misfortune to have come out with *Engines of Creation* just before the moment of paradigm crisis, a few years in advance of the turning point, which, if it had to be located in time, occurred in 1990, the year of the IBM thirty-five-atom logo. The tectonic plates had started moving long before that, of course, with Feynman's "Room at the Bottom" — which, at the time, people treated mostly as a joke. Hans Dehmelt was part of the shift, what with his pet particles Astrid and Priscilla, and his cranky mumblings about "a still persisting wave of quantum-mystification in the literature." The STM was part of it, too, as was IBM's one-atom switch — not to mention all the atomic names, acronyms, dumb pictures, and other assorted molecular stunts.

What all of these modern miracles proved was that atoms were no longer the invisible nameless nothings of yore. They were in fact, and could be treated as, mechanical, Newtonian entities. They were *objects* that could be made to *do* things.

Still, it was a far cry from that realization to the further conclusion that you could grab on to those atoms one by one and join them together so as to build an actual machine — an actual working system — one that (with the proper programming) could construct anything that was physically possible, automatically and essentially for free, meanwhile providing you with surplus electrical energy as a by-product.

That was a lot to swallow all at once, and few were able to. It was only natural to think, therefore, that you really didn't have to pay all that much attention to Drexler's arguments, or to his printed replies to likely objections. Why bother reading all of his five hundred pages of dense text — plus equations — if you knew in advance that it had to be wrong? A short, reassuring glance would be enough. Any slight difficulty, any doubt, was enough to prove that the whole colossal edifice rested on quicksand.

Even so, when some of the world's prime scientific authorities took the time to look at Eric's molecular machines and devices, what a shock it was when they actually pronounced them sane! In fact it turned out that for every chemist who said that one of Drexler's proposed structures would explode on contact or wouldn't work or couldn't be built, there was another of equal stature and authority who rendered the precise opposite judgment.

Leo Paquette, the Ohio State University chemist who'd synthesized dodecahedrane (a complicated organic molecule), examined Drexler's "overlap repulsion bearing" and his "strained-shell sleeve bearing" (the Michelin-tire things), and even his 3,557-atom planetary gear, and said: "If the molecules could be synthesized (and this is certainly *not* the case at present), they would be quite stable entities. I have no doubt that his structures will be *entirely stable.* No tendency for explosion would be contained in these materials once constructed."

And Roald Hoffman, Cornell University's Nobel laureate in chemistry, inspected the same Michelin-tire molecular configurations and said: "I think the structures are likely to be stable. I think it is likely that the rotation will be smooth. The gearing I'm less certain about — because I think there may be some flexibility or softness in the structures; they may deform under strain."

As for the extremely unhappy "sheets of carbenes" that Eric supposedly wanted to make use of, the fact was that Drexler had never even talked about them. There were no "carbene sheets" anywhere in *Nanosystems,* or in the doctoral dissertation on which it was based.

"To my knowledge," said Rick Danheiser, who'd been on his examining committee, "Eric has never discussed 'carbene sheets.' "

"I know how someone could get that impression, though," said Drexler. "Namely through a careless misreading of figure 8.18 in *Nanosystems,* as well as the associated text."

Figure 8.18 showed a diamond surface; the associated text said: "Simple truncation of the bulk structure would yield a surface like that shown in Figure 8.18(a), covered with carbene sites."

But a truncated structure — such as was created by cutting a stick of butter with a knife — gave you a *surface,* not a "sheet."

And if his structures were dull, uninteresting, retrogressive, and medieval, Drexler was eager to see more advanced designs.

Go ahead! Just try to design one!

As for Rick Danheiser's complaint that Drexler was using inherently unstable materials in his designs, Drexler's attitude was that he used such designs intentionally and advisedly, precisely because — as Danheiser recognized — such designs "had this or that desirable property." Why *not* take advantage of those properties, assuming that you could actually construct the object? If future experience proved that you couldn't, well, *that* was when you changed the design, not beforehand.

As for Whitesides's and Danheiser's claims that nobody knew how to *make* Eric's molecular structures — well, others were not quite so pessimistic.

"It is hard to imagine that such structures could become readily available in less than a decade," said Clark Still, the Columbia University chemist. "It would just likely take a very considerable effort."

Less than a decade? Even Eric himself was never *that* hopeful.

And as for Danheiser's contention that "one might just as well develop designs for interstellar spacecraft," the fact of the matter was that the world's first interstellar spacecraft had been designed, built, and launched a decade before. On June 13, 1983, indeed, the *Pioneer 10* spacecraft (carrying its "interstellar plaque") had departed the known solar system.

And, finally, where J. Fraser Stoddart had claimed that the structures pictured in *Nanosystems* were "bizarre," and that Drexler's knowledge of chemistry was "scant," the fact was, as Stoddart admitted: "I haven't read the book, to be honest."

Haven't read the book?

"Probably the best student I ever had, he took it — he's now in Zurich — he took it and did a group meeting on it. And we just felt that when it came to the chemistry it was a bit bizarre, really. Things that are being talked about would not be worth doing, and could not be done, so you don't really enter the realm of reality, you just sort of live in a foggy land of gibberish.

"Far be it from me to say things cannot be done and will not be done," he added, "but I live on the real planet, as far as chemistry is concerned. When I look at chemical structures I get an immediate feel in seconds that, oh yes, that's reasonable, that's possible, or that's just, you know, crazy."

The student who actually "did the group meeting" was Douglas Philp, who was now at the Federal Institute of Technology, in Zurich. He, at least, *had* read *Nanosystems.*

"I read his book," Philp said. "I presented a sort of overview of the field of nanotechnology and what I saw as the problems with Eric Drexler's approach.

"In principle everything's fine. But when you actually get down to look at the chemistry that he proposes, it's completely ridiculous. I mean there's no real basis for it in reality. A lot of it has a very classical approach to mechanics, the actual physics of it."

There was a basic distinction, however, between chemical

structures that were "ridiculous" in the sense they couldn't be built now, because the techniques for making them didn't yet exist, versus structures that were ridiculous because they violated the known rules of chemistry. Had Drexler proposed any designs that were ridiculous in that sense?

"There's none that I saw that violated fundamental rules of bonding," said Philp. "But if you'd give them to a graduate student to synthesize, you'd probably be told where to get off, really."

In the end, it was hard to shake the impression that Drexler's two major mistakes were, one, proposing structures that can't yet be built, and, two, failing to consult with the chemists who would have informed him of this fact.

"He needs to relate more closely to chemists and talk with them and collaborate with them," said Fraser Stoddart. "Then I think if he uses chemical structures, he may produce structures in his articles that relate to chemistry. At the moment, as I recall from looking at the book at the time, within ten seconds you're saying, 'Now wait, this can't be for real!' "

"You know what Eric should do?" Julius Rebek said. "He should sit down with a practicing chemist and present examples that could be within our grasp, rather than make up some really absurd-looking molecule."

Five days before *Nanosystems* was officially published, Drexler had written a short note that said: "I predict that the last refuge of critics will be to deny that anything short of a physical demonstration can provide solid evidence for the feasibility of something new like molecular manufacturing. This position, of course, denies the usefulness of all but the most routine sorts of engineering, and comes close to denying the value of any kind of scientific theory. It can be called the 'wake me up when it's over' school of criticism."

His forecast had been proven true enough. From his own view, at least, he'd done everything necessary to show that molecular machines could be constructed, that they were chemically stable and physically possible, and that they'd operate more or less as described. One thing he'd made a special point of in *Nanosystems* was to anticipate objections and answer them in advance.

His chapter on "positional uncertainty," for example, pretty much laid to rest the recurrent objection — regular as the sunrise — that Brownian motion, thermal vibration, *kT,* would play havoc with his molecular machines. That single chapter had a hundred separate equations in it — exactly one hundred — plus logarithmic plots showing what the actual physical dimensions were of the vibration to which molecules of various types were subject throughout a whole range of temperatures. There was such a mass of technical information there, *so* much physics, that those who bothered to look at it, and who understood it, gave Drexler no more flak about "the thermal vibration problem."

But there was a commonsense answer to the whole thing, too, which anyone could follow. For one thing, although life was hell in Brownian motion, such large-scale vibratory motions actually existed only in liquids; in solids molecules merely "jiggled in place." And while the "exemplar manufacturing system" — the desktop factory, the black box — would use liquids for raw materials, the normal Brownian motion of its feedstock particles presented no problem while they were in the holding tank. The molecules were only waiting to be processed, so who cared what they did at that point?

The manufacturing unit itself and all the parts of which it was composed, by contrast, would be solid objects, whose molecules jiggled in place. But such jiggling presented no problem: in most cases the magnitude of the vibration would be too small to matter, and in the few cases where it did matter all you'd have to do was to make the affected part larger, or stiffer, or both — thus lessening the effect of the molecular vibrations. You'd also design the structure so that each molecule's chemical bond to the next one was strong enough to withstand the vibrations that remained, so on and so forth. There were a number of ways around the problem.

So far as Eric Drexler could see, there was absolutely no obstacle, no barrier, no bottleneck, standing in the way of his "exemplar manufacturing-system architecture." The molecular cornucopia was a real prospect. Everyone could have their own little black box.

And now if only people would start working on it, from the bottom up.

15

"Good Luck Stopping It"

In November of 1993, Rice University announced a "nanotechnology initiative." The idea was to put up a new building on the campus in Houston and populate it with nano-inclined experts from various fields and departments. Here researchers would create the founding works of the new realm: the molecules, the structures, the nanomachines of the future.

Prime force behind the initiative was one Richard E. Smalley, codiscoverer of C_{60}, also known as the buckyball. Smalley, it must be said, was firmly located on the other side of the paradigm shift. For one thing, he was that rarest of all birds in academic circles, a confessed admirer of Eric Drexler.

"I'm a fan of Eric," he said. "I am a fan of his, and in fact in my endeavors to explain to people what I thought the future was, particularly the board of governors here at Rice, I have given them copies of some of Eric's books."

That future, in Smalley's view, included nanotechnology in a fundamental way.

"Science and technology on the nanometer scale is very likely to be one of the most important technologies of the twenty-first century. It may even be the most important. After having studied it a bit on the university-wide level, we've found that it has a remarkable synergy aspect between various departments and disciplines, in fact crossing the engineering–science barrier beautifully. And because of that we find it a very useful rallying flag to direct our thoughts to future recruitment, building instrument centers, and ultimately, we hope, in the education of the university. Why should we be teaching students to become scientists and engineers in the old technology? They should be part of the future.

"And so what has happened is that we are going to put up a new building here which will house about half the chemistry department, but will not be a chemistry building. In fact Rice will not *have* a chemistry building. The building will be named after its principal donor, whoever that turns out to be, but it is being called around here the Nanotechnology Building."

Smalley, like Drexler, was decidedly of the engineering persuasion, having grown up during the *Sputnik* years. "The most romantic thing you could possibly be in those days was a scientist or engineer. This was where the action was."

He took a job at Shell Chemical, in New Jersey, where he got hooked on chemistry; ultimately he took a Ph.D. in the subject from Princeton. In 1985, now at Rice, he and some colleagues placed a small bit of graphite inside a laser vaporization apparatus and discovered that they'd created a strange new form of carbon.

Carbon, the sixth element of the periodic table, was known to occur naturally in the form of "network solids" such as graphite and diamond. In both of those forms, each carbon atom was connected to four others, and each of those to four more, and so on, in large, spread-out networks. In graphite, these networks ran in flat sheets, the layers of which slid across each other easily. In diamond, by contrast, the atoms were ordered in rigid three-dimensional cubes, the arrangement that gave diamond its hardness. For years it was thought that this was the only way in which carbon came: in long-drawn-out continuous systems.

But when Smalley and cohorts zapped some graphite in their super-duper laser-beam gadget, they got a bunch of tiny carbon marbles instead, a hitherto unknown form of the element. Sixty separate carbon atoms had somehow gotten together and joined up to compose a discrete and self-contained molecule, a tiny, hollow sphere. Further examination revealed that the sphere had a soccer-ball-like shape, consisting of thirty-two faces: twelve pentagons and twenty hexagons organized in such a way as to make up a so-called truncated icosahedron, one of the Archimedean solids known since antiquity. Smalley and crew named the molecule "buck-minsterfullerene" ("buckyball," for short), after the geodesic domes of Buckminster Fuller, which they closely resembled.

For several reasons, the buckyball (chemical designation: C_{60}) caused a mania among working chemists. For one thing, the molecule had an undeniable aesthetic appeal: mathematically, it was the most symmetrical molecule that was physically possible. "It is literally the roundest of round molecules," said Smalley, "the most symmetric molecule possible in three-dimensional Euclidean space."

Second, buckyballs gave rise to some extremely unusual electrical behavior. Depending on how C_{60} was mixed together ("doped") with other substances, it could function as an insulator, a conductor, a semiconductor, or a superconductor. By any measure, that was a lot of ways for one and the same molecule to operate.

Third, because it was a hollow, open structure, C_{60} allowed other atoms to be trapped, or "caged," inside it. Accordingly, chemists now placed atoms of various elements — potassium, cesium, and even uranium — inside buckyballs, and gleefully spoke of "shrink-wrapping an atom."

The buckyball was a grand new toy in the chemists' playpen, one on which they lavished untold amounts of "research," generating some fourteen hundred scientific papers about it and related fullerenes in the space of a few years. "We're like kids who have just discovered Tinkertoys," said Donald Huffman, of the University of Arizona.

In what had by now become a familiar pattern, the scientists who were exploring this new field spoke elliptically of the countless "uses" to which their playthings could be put, without, however, actually mentioning any.

"I'm just sure that years from now we're going to be seeing all kinds of new materials and practical applications," said Huffman. "It could turn out there are a lot of things we get from this stuff."

"We do not know what the fullerenes' burgeoning traits will allow," said another researcher, "but it would be surprising if the possibilities are not wonderful."

Rick Smalley, likewise, wanted his "babies" to do real work. "What I want most is to see that *x* number of years down the road, some of these babies are off doing some good things."

Smalley was much excited, then, by the addition of the buckytube (a single-walled carbon pipe, also called a "nanotube") to the ranks of fullerenes. Buckytubes, of course, were anticipated to have all sorts of fantastic nameless applications, but Smalley himself actually came up with one: the "nanofinger," a long, slender rod with which to move atoms. Put two such rods together, like tweezers, and you'd have yourself an atomic "hand."

"A great milestone would be to get two nanofingers together so you can pick something up," he said. "So far the image you get of the STM and these local-probe things is that you've got a finger and you're moving it around on a table and moving it up and down. In fact a better analogy is your elbow, something that's not long and skinny but is big and fat. It's really difficult to go scanning around the room and report what's there if you've got something as coarse as your elbow to detect what's going on. Even so, it would be nice to get two elbows together.

"I think that a buckytube being the probe of an STM would be a help, probably even qualitatively a help, but the big breakthrough will be to get two of them so you can oppose them, like Chinese chopsticks. It's like the development of the opposing thumb, to pick things up. We have no way right now of picking something up by holding it between two things. The opposing thumb would be a major advance."

Conceivably, it would take us one step closer to the Great Nano Future as envisioned by Eric Drexler.

"What I like about Eric is his very vivid imagining of how exciting this future could be," said Smalley. "I'm very serious about this. I'd love to build one of these things, so when I read what he's talking about, I'm ready to do it.

"Somebody's gonna do this someplace. I'd dearly love it to be us."

* * *

Smalley's was by no means the only organized nanotech effort in existence, but it was one of the most narrowly focused on the kind of thing that Drexler had been talking about for nigh onto twenty years now. Many of the others, as it happened, were in Japan.

The Japanese, as was no longer news, were at the leading edge of technology development, especially when it came to miniaturization. It was no big surprise, then, when in 1991, MITI, Japan's Ministry of International Trade and Industry, announced a $200 million project "to promote research into nanotechnology."

Japan, indeed, boasted an incredible number of laboratories and projects devoted to something that at least sounded like nanotechnology. There was, for example, a Lab for Nano-Electronics Materials, a Lab for Nano-Photonics Materials, and a Lab for Exotic Nano-Materials. There was the Yoshida Nanomechanism Project, the Hotani Molecular Dynamic Assembly Project, the Kunitake Molecular Architecture Project, the Nagayama Protein Array Project, the Aono Atomcraft Project, and so on and so forth.

Whatever the reason, the Japanese seemed to be less straitjacketed by outmoded paradigms or other fixed preconceptions that acted (among those afflicted by them) as mental roadblocks to thought and action. Anyway, Eric would go over there every year or so, to give lectures, and despite the language barrier and everything else, he got a better reception in Japan than he normally got in the United States or Europe. He'd even been offered two visiting professorships at Japanese universities, both of which he declined.

Still, there was the problem of knowing exactly what was going on in the various Japanese labs and projects, because as was becoming increasingly clear, not everything that was called nanotechnology was in fact nanotechnology in Drexler's sense of the term. In Japan, for example, some of the money allegedly earmarked for "nanotechnology" research was actually being spent on tiny robotic pipe scrubbers with which Japanese power companies hoped to clean out their electrical conduits. That was microtechnology, not nanotechnology.

For better or worse, the latter term had developed a cachet of its own and was now being extended to include all sorts of phenomena that by rights had no business being called by that name. Of course, Eric hated what he regarded as this rank bastardization of his nomenclature, particularly the word that he'd always regarded himself as having coined. The term *nanotechnology*, he said, was now being used "to glamorize the production of nanoscale blobs."

Which was true. A group at Harvard was extremely proud of having "nanoengineered" a bunch of tiny little water droplets, each one a mere 10^{-7} μL (microliters) in volume. "No one has made water droplets that small before," they said. ("Producing cigarette smoke would also be 'nanotechnology' by this criterion," said Drexler.)

In fact the term was being used to glamorize far more than nanoscale blobs: there was now a Nano shampoo and a Nano hair conditioner on the market. (For that matter there was also on the market "DNA" perfume, by Bijan Fragrances of Beverly Hills, plus a book called *Quantum Golf.*) There were materials companies called Nanophase Technologies and Nanodyne. And off in its own separate corner there was Nanothinc, a California nanodevelopment firm that included three nanosubdivisions: Nanocomm, Nanoventures, and Nanotainment.

(Nanotainment.)

The greater universe of business and finance, apparently, never suffered any major embarrassments over paradigm shifts.

In an attempt to stanch this flood of imitators, clones, and loathsome free-riders, Drexler now coined another new term: "*molecular* nanotechnology." "With luck," he said, "the term will prove bulky and awkward enough to retain a distinct meaning."

But who could blame people for jumping on the nano bandwagon? Two of the biggest technologies in the last quarter-century, after all, were ones that operated at the tiniest scales — biotechnology and computers — and there was absolutely no mystery as to why this was so. The molecular level was where the action was, it was where the properties of all larger objects were established and determined. It was the difference of only a few invisible nucleotide bases of a DNA molecule, for example, that determined whether a newly fertilized cell became an ant or an elephant, a mouse or a whale.

Just a few different molecular groups down there at the bottom and you got this enormous, truly staggering difference at the top, in end result. The secret was that each of those tiny differences got repeated and distributed in the act of cell replication: each one was augmented and amplified millions of times over. Plus the cells themselves developed as they went, so that a daughter cell was not always an exact copy of the parent cell, but was subtly dissimilar, allowing for the differentiation of the new cells into all of the body's various shapes and structures.

So the tiniest invisible change at the bottom could have the largest and most palpable effects at the top, where people actually had their being and led their lives. Obtaining control of that realm would therefore give you the biggest possible leverage or mechanical advantage: the greatest gain in output per unit of input. That was where nanotechnology's power came from, that was the origin of its magical force.

Nanotechnology, though, was not the only attempt to cash in on the leverage generated by working at the source, at the very root, of the structure and complexity of the Big World, although it was without a doubt the most ambitious and the most far-reaching in its consequences. But the other such attempts were portentous enough.

There was the Human Genome Project, for example, the purpose of which was to map out the complete molecular blueprint — the whole long sequence of base-pairs, all three billion of them — of a human being. This was not just "knowledge for knowledge's sake"; rather there was a practical purpose behind it: access to the blueprint would allow you to make alterations to the end result. Despite the way it sounded, this was not a question of "changing human nature" but rather of correcting a naturally induced deviation from the norm.

Many diseases had a genetic origin, meaning that they stemmed from a stretch of the DNA molecule where the base-pairs, or nucleotides, were not in the normal sequence. Sickle-cell anemia was the most famous example: it resulted from the incorrect placement of a single DNA base-pair on an otherwise intact chromosome. That was "leverage" of the worst sort: some misplaced molecules and the person died. But the converse was that if you could put those stray molecules back where they belonged, then that person would be cured of the disease.

That was the idea behind gene therapy. The theory was that if you could locate the DNA sequences that were responsible for a given disease, then you could treat the illness simply by inserting the right sequences. You'd physically rework the DNA molecule and send the patient home cured.

Well, this was unbelievable, it was impossible, it sounded like magic — and so of course right on schedule, precisely as expected, some of the day's most distinguished and enlightened scientists took time out to express a reservation or two about the prospects of gene therapy ever actually happening.

In 1971, Sir Frank MacFarlane Burnet (who'd shared the 1960 Nobel Prize in physiology or medicine) said: "I should be willing to state in any company that the chance of doing this will remain infinitely small to the last syllable of recorded time."

And in 1977, Ernst Chain (who'd shared the 1945 Nobel Prize in physiology or medicine) said: "There exists no method, nor is there the likelihood that one will be discovered in the foreseeable future, by which it would be possible to alter the nucleotide sequence, and thereby the genetic properties, in any gene of a mammalian cell in a controlled manner which could be called 'genetic engineering.' Any speculations that such a process may be near at hand and could influence the heredity of man must be dismissed as science fiction."

Science fiction! Last syllable of recorded time!

(Modest utterances of this sort once prompted Allan Sandage, the astronomer, to say that in cosmology, "progress in this field only comes at the funerals for the astronomers." That, apparently, was where your paradigms actually shifted: at the funerals.)

This latest case of "science fiction" turned into science fact in 1990 when two young Ohio girls suffering from the immune disorder known as "bubble-boy disease" (named after the boy who lived most of his short life in a plastic bubble to protect him from infection) were successfully treated by being injected with new genes.

The disease in question, formally known as ADA (adenosine deaminase) deficiency, was caused by some out-of-place nucleotides on the DNA of chromosome 20. To insert the right ones, the plan was to use a "viral vector" as the delivery medium.

A viral vector was a virus that contained the right molecular sequences and which furthermore had the natural inclination to insert those sequences into other cells. Supposedly, the vector would carry the right nucleotides to the impaired section of the DNA, insert them, and then vanish again for parts unknown. (It was precisely this scenario, using viruses to transport new genes into ailing cells, whose probability MacFarlane Burnet had said would be "infinitely small to the last syllable of recorded time.")

After this sort of "reverse-infection" process, the patient would emerge good as new — indeed *better* than new inasmuch as she'd now be cured of a malady that had been in her very genes from birth. If it worked, it would be the world's first case of a virus not causing a human disease, but curing it, all under human direction and control.

And in 1990 it worked, when three physicians at the National Institutes of Health (W. French Anderson, Michael Blaese, and Kenneth Culver) extracted white blood cells from the Ohio girls, combined those cells with the viral vectors, and then injected the "corrected" cells back into the two patients. The transformed cells started churning out the normal amount of ADA, precisely as if they'd been equipped with the right molecular sequences to start with. This same sort of treatment, the scientists thought, could cure every known form of genetic disorder, from Alzheimer's disease to cancer.

Gene therapy was not, of course, molecular nanotechnology in Eric Drexler's sense of the term: it was not molecular manufacturing, not an example of mechanical assemblers moving molecules around and building up an array of useful objects. But it *was* a form of molecular technology, a case of manipulating molecules deliberately and for a purpose, gaining leverage over the Big World by fiddling with the nanoscale entities of which it was made.

It was a regular fixture of Drexler's overall view of things that canonical nanotechnology was a virtually inevitable technical development. This was not merely an impression or a vague hope on his part — far from it. He'd worked out an entire argument, a proof, that the nanotech revolution could hardly be avoided. The argument presented a stepwise, pincers-type case for its conclusion, from which, indeed, there seemed to be no easy escape.

Premise one was the "multiple pathways" point, the observation that there were many distinct alternative routes, all of them leading toward the fabled assemblers. There was, for example, the protein-engineering pathway, where you'd construct an amino acid sequence that would fold up into the shape of a molecular machine, or at least into the shape of a lesser component. But that was only one possible route: there were now a bunch of proximal probes — STMs, AFMs, and so on — with which you could position atoms exactly where you wanted them, to gain the same result. And finally there was the self-assembly route, a third distinct road into molecular machine territory, where you'd design molecules to have shapes and bonding sites such that they'd fit together precisely, lock-and-key fashion, so as to produce a functional device of some sort.

The advantage of having multiple pathways, of course, was that a roadblock along any one route still left all the others open. That, in turn, meant there was no single place at which the forward march to the assemblers could be stopped.

Premise two of the argument was that there were rewards and payoffs at every step of the way, along each of the various routes. And you could reap those rewards, in the form of increased capabilities and wider options, quite aside from the goal of creating a molecular manufacturing technology. You didn't have to be a believer in the greater Drexlerian vision, in other words, in order for progress in your field to help not only you, but nanotechnology as well.

And then finally, when those various techniques and pathways had been refined to the point where the goal of molecular manufacturing was actually within reach, nanotechnology would appear as a huge and attainable boon; this was premise three. At that stage, when the whole nano dream seemed to be poised on the brink, then people would actually do it: they'd go ahead and create the technology. The lure of it all would be too powerful to resist, especially in view of world competition, which would be a prime driver of the whole enterprise.

"I like to compete," said Rick Smalley. "I like to be on the team that did it first."

He wasn't the only one, and in fact each of Drexler's "multiple pathways" was soon marked with a row of new nano-milestones. Du Pont's protein-design team, for one, went on to make a new-and-improved artificial protein, a structure that would act more like a usable device than the earlier version was likely to.

Early on in the game, Bill DeGrado had noticed that his original, made-from-scratch α_4 protein was not exactly the picture of stability he'd wanted it to be. Its component helixes jiggled and slipped like a bundle of worms, all because of the fact that the components were not held together by fixed bonds. The obvious solution was to introduce some sort of molecular cross-bracing: insert a few metal atoms, for example, as stabilizers. By 1993 DeGrado and his colleagues had made the change, and their new protein (called H6-α_4) featured zinc-atom cross-struts that held all four helixes stiffly in place.

Other researchers had come up with ways of incorporating "unnatural" amino acids into proteins. The proteins of all living things were combinations of naturally occurring amino acids; but there were only twenty different naturally occurring amino acids, whereas some sixty or more were chemically possible.

Soon enough, scientists had figured out ways of getting the cells to produce "unnatural" amino acids. They wouldn't be made in test tubes: the cells would make them, turning out substances they were never meant to build.

It was wonderful news to nanofans, who saw a whole new range of possibilities opening up in front of them.

"In a fully artificial system, the twenty natural amino acids might even be entirely dispensable," said Greg Fahy, a longtime nano enthusiast. "This could be expanded to include an unlimited number of unnatural amino acids, and these unnatural amino acids could contain totally nonbiological catalytic groups or even premade machine parts, such as structural support struts, molecular bearings, or the like."

But if nature's amino acids were entirely dispensable, then why not the proteins themselves? Maybe you could have a protein substitute, a mock protein, an alternate material that you could engineer and fool with to your heart's content.

Drexler had proposed that, too, back in his 1981 *PNAS* piece, where he'd said that in the wholly unlikely event that all attempts at protein design failed, there was still the option of creating, from scratch, an equivalent material that might be more amenable to deliberate engineering. "If protein design were to prove intractable (because of difficulties in predicting conformations), this would in no way preclude developing an alternative polymer system with predictable coiling and using it as a basis for further development."

Now, any calm and skeptical reader might have seen this as whistling in the dark — as yet one more case of "science fiction" being gussied up as science — but a few years after Drexler had predicted it, even that rather farfetched item had actually been invented. In 1993 researchers at the University of California at Berkeley created an analog polypeptide (a substitute protein, essentially), called an oligocarbamate. This new substance had a molecular backbone and side chains just like conventional proteins did, but it was made out of slightly different materials, ones which had the added attraction of being both stiff and highly controllable. In short order, the inventors had developed a "library" of some 256 oligocarbamate structures.

So now there were substitute proteins to work with.

A whole new pathway!

And then, suddenly, there was the "artificial atom."

This was so bizarre an invention — and it was an "invention"—

that neither Drexler himself, nor, very probably, anyone else, had ever even remotely anticipated it. How could there be an artificial *atom?* But Raymond Ashoori, a physicist at AT&T Bell Labs, had created one, and there it was whether you liked it or not — an atom whose electron count was controllable by its human maker, from zero to sixty.

The "atom" in question was actually an empty space within a gallium arsenide crystal to which electrons could be moved one at a time by the application of a light magnetic pulse. In the case of an ordinary, garden-variety atom, electrons were held in place by the nucleus, whose positive charge attracted and bound the negatively charged electrons. In an "artificial atom," by contrast, electrons were held, instead, by an externally imposed magnetic field. But the final effect was much the same: a bunch of electrons whizzing around in a small space.

"By observing how much energy it takes to add each successive electron, we can directly learn how the electrons interact with one another," said Ashoori. "One of the most exciting effects is seen with just two electrons in the atom. We are able to observe the signature of a single electron 'flipping' in response to an applied magnetic field."

Best of all, you could use this artificial-atom generator — this "toy box," as Ashoori called it — you could use it to design your own atoms.

"We can make atoms of any size," he said. Horst Stormer, Ashoori's coworker at AT&T, added: "You can make any kind of artificial atom — long, thin atoms and big, round atoms."

The amazing conclusion, of course, was that maybe you could link some of those newly created atoms together, thereby creating your own artificial molecule. And then maybe you could join those artificial molecules together to produce — why not? — an artificial solid.

Was this not the blunt future already staring us smack in the face? Here were two staid and serious corporate physicists — practical men, laboratory types — here they were talking about "real" versus "artificial" *atoms*, the bright new amusements they'd made in their little "toy box."

It was a strange and wondrous sight. The twentieth century had passed through paradigms as if they were the staged phases of child development — it had gone from there being no proof of atoms, to atoms as real but unseeable; then to atoms as real, seeable, and even separately movable; and then, finally, to do-it-yourself, customized (skinny, medium, or full-figured), artificial atomic creations.

No longer was the human race confined to the "building blocks of matter" that Mother Nature had provided; now we could make our own.

Compared to which, nanotechnology was not all that outlandish a prospect. Nanotechnology, after all, used only nature's atoms, *normal* atoms, the tiny marbles that during these latter-twentieth-century days had been individually touched, pushed around, lifted and lowered, played with, bottled up, treated as pets, and given their own names.

All nanotechnology wanted to do was to take those same objects and organize them into working machines.

Was that so crazy?

By mid-1994, the atomic realm had been colonized by a weird assortment of man-made shapes, structures, materials, and devices. There were now in existence any number of letters, words, logos, and pictures, plus various other contrived molecular objects, each of them composed of small and countable numbers of atoms. There were, in addition, atomic switches, self-replicating molecules, and molecular shuttles and trains. There were buckyballs, nanotubes, atomic corrals, nanowires, and molecular propellers and gears. There were deliberately engineered artificial proteins, *faux*-proteins, and "unnatural" amino acids.

There were even "artificial atoms" . . . in the better class of laboratory.

These atomic structures were only the bits and pieces of something greater, the mere makings — comparable to a bunch of unrelated nuts and bolts scattered across a shop workbench. But the important fact was that they were there at all, these atomic parts and appendages, because each of them had been deliberately designed and engineered, and then crafted into existence. They hadn't just dropped out of the sky like snowflakes: they'd been planned and premeditated, and then actually produced.

Now if only someone would just *put them all together* . . .

And it seemed to be only a matter of time before someone in fact would. A year after the publication of *Nanosystems*, more than nine thousand copies of the book had been sold. For a text whose avowed purpose was to demonstrate that "an exemplar manufacturing-system architecture" was in fact possible — a kitchen appliance with which you could order up anything in a vast catalog and then watch it plop out an hour or so later — that was a lot of copies. They weren't being snatched up by Star-Trekkies — not this technical text with 450 equations in it. They were finding their way to the right audience: applied physicists, experimental chemists, people doing lab work.

And so it seemed to be only a matter of time before the unrelated atomic bits and pieces coming out of the laboratories collided with the specifics of Eric Drexler's master plan, out of which fateful confrontation would emerge a bunch of nanocomputers, a covey of molecular assemblers, and then, finally, the little black box itself: the automatic, programmable device that would build for you, for next to nothing, everything you ever wanted.

As to who would be the first to do it — the where and the when — that was something no one knew. What Drexler did know, or at least strongly believed, was that sooner or later molecular nanotechnology would actually burst upon the scene, and probably sooner rather than later.

And who could deny it? When he made a prediction in his chosen field, he was very seldom wrong.

Back in the Cambridge days, when all of this was still an idea, just a hope, the actual outcome was highly ambiguous and uncertain: maybe it would happen someday or maybe it wouldn't. And in fact even much later, during the golden age of the NSG, it appeared that you could still prevent nanotechnology from happening, head it off at the pass, just by keeping quiet about it. Some of the kids had even joked that what the letters *NSG* really stood for was "Nanotechnology *Suppression* Group."

But suppression was no longer in the cards. Too many people

in too many research labs in too many countries had made too many advances for nanotechnology to be staved off forever.

These days, whenever Drexler heard someone arguing against nanotechnology, saying that it shouldn't be developed because of the changes it would bring, because of the dangers, because it was "scary," the same thought always popped into his head: *Good luck stopping it.*

part iii

Paying the Piper

16

Sunday Afternoon in the American Suburbs

A hundred years ago, in 1895, H. G. Wells published a short novel called *The Time Machine.* It was the story of a man who traveled into the far future, to the year A.D. 802,701, and returned to tell his friends what lay in wait off in the distance, beyond the event horizon.

Largely, it was a pretty picture. Planet earth had become a garden.

> The air was free from gnats, the earth from weeds or fungi; everywhere were fruits and sweet and delightful flowers; brilliant butterflies flew hither and thither. The ideal of preventive medicine was attained. Diseases had been stamped out. . . . One triumph of a united humanity over Nature had followed another. Things that are now mere dreams had become projects deliberately put in hand and carried forward. . . .
>
> Social triumphs, too, had been effected. I saw mankind housed in splendid shelters, gloriously clothed, and as yet I had found them engaged in no toil. There were no signs of struggle, neither social nor economical struggle. The shop, the advertisement, traffic, all that commerce

which constitutes the body of our world, was gone. It was . . . a social paradise.

People lived in peace and harmony, in a condition of perfect health and perpetual youth: "aged and infirm among this people there were none." These childlike and simple men and women put flowers in their hair, danced and sang, and gamboled about in the sunshine. "They spent all their time in playing gently, in bathing in the river, in making love in a half-playful fashion, in eating fruit and sleeping." Life was a mixture of art, play, eroticism, and sleep — a "contented inactivity."

There had been some changes, of course, in the physical constitution of these people. The sexes, for example, had grown more alike — which, when the Time Traveller thought about it, was not at all surprising. "This close resemblance of the sexes was after all what one would expect; for the strength of a man and the softness of a woman, the institution of the family, and the differentiation of occupations are mere militant necessities of an age of physical force."

And there had been changes, too, in their psychological makeup. They were less strong and quick, and less intelligent, than the human beings of bygone days. Not that this was any great mystery: it was "a logical consequence enough," the Time Traveller told his friends. "Strength is the outcome of need; security sets a premium on feebleness. . . . It is a law of Nature we overlook, that intellectual versatility is the compensation for change, danger, and trouble."

But in the world of the far future, danger and trouble were things of the past. People lived in "almost absolute safety," had "no need of toil," and wound up as the contented cows of the cosmos. "Very pleasant was their day, as pleasant as the day of the cattle in the field. Like the cattle, they knew of no enemies and provided against no needs."

They were "humanity on the wane"; it was "the sunset of mankind."

"I grieved to think how brief the dream of the human intellect had been," said the Time Traveller. "It had committed suicide. It had set itself steadfastly towards comfort and ease, a balanced society with security and permanency as its watchword, it had attained its hopes — to come to this at last . . . a perfect conquest of Nature," a "too perfect triumph of man."

It was the start of a long line of science fiction stories about "the future." Many of them, for whatever reason, were built around the identical theme: that no matter how wonderful the future appeared in prospect, it would actually turn out to be a catastrophe.

The major remaining question for nanotechnology was whether it was a project worth pursuing. Nanotechnology would

K. Eric Drexler. (*Brent Pederson*)

give you, as Drexler had said, "effectively complete control of the structure of matter" — or as Rick Smalley had put it, "as much control as you're going to get."

But was such control worth having?

It would allow the human race to do every last thing that was physically possible; it would let us advance toward the limits established not by temporary conditions on earth (*Limits to Growth*–style) but established by nature, by the very laws of the cosmos.

But was reaching those limits a good thing? Was it a net gain, or a net loss to humankind? What would happen to the frontier, what would be left to do, once every possible deed had already been done?

These were not questions that Eric Drexler spent any large amounts of time worrying about. He pretty much took it for granted that having complete control of the structure of matter was a fine thing, that reaching "the limits of the possible" was a blessing to mankind.

But they were questions worth asking. They were not scientific or technological issues, of course, but rather ones involving human values: they concerned ideals and hopes, one's aspirations for the species. They concerned assumptions about human nature, about what was the morally proper and fitting goal of human life.

Answering those questions in any satisfactory fashion required making a forecast, a prediction of the likely consequences of nanotechnology, both for individuals and for society. And then it required making a judgment whether those consequences were, in the end, harmful or beneficial.

When Drexler considered the subject of "consequences," he tended to think in terms of physical risks, the threat of which had kept him mum about nanotechnology for three or four years after he'd first gotten the idea. Later he spent inordinate amounts of time trying to come up with strategies for avoiding the evil that people could do with an army of nanohelpers at their disposal.

The fact remained, though, that the physical risks of nanotechnology might not be the worst ones of all. Far more serious might be those that were social and psychological.

Physical dangers, after all, were nothing new to the human

race. People had always had to contend with them in one form or another. They were the price you paid for living: not only sickness, aging, and death — the usual joys of human existence — but also war and crime, mayhem and mischief of every description. These things were *normal.* They were ordinary and expected. They were the very stuff of life.

And of course precisely because they were so omnipresent and everyday, people had evolved all sorts of tried-and-true (if ultimately ineffective) methods for dealing with these routine threats to life and limb. Drexler's strategies were of a piece: you physically isolated the "bad" assemblers or you incapacitated them so that they couldn't operate outside the lab; you fought the "bad" assemblers with "good" assemblers, down in some invisible nanoscale war zone; and so on. All of which were entirely logical responses: they were what people had done from time immemorial; they were nothing new under the sun.

But having complete control of the structure of matter *was.* Nobody'd ever had to cope with that before. Far more frightening, definitely more paralyzing to the imagination, than the sight of nanotechnology going wrong was the prospect of its going *right,* of its control over matter being all too complete.

So the primary philosophical question was left dangling: Was nanotechnology worth doing? Could people handle the largesse of it all? The abundance, the bounty . . . the boredom?

There were no computer modeling programs, unfortunately, to help out with those problems. Scientists and engineers, likewise, were not quite adequate to the situation. This was the province of the humanist: the philosopher, the psychiatrist, the cultural anthropologist.

But when those thinkers considered the question, or tried to, no one clear answer emerged.

"The idea that there is going to be the raw materials needed for doing all of this stuff, and that they will be distributed, and that it will be financially, economically feasible to have these little buggers do the work, I'm really not sure at all," said Mihaly

Csikszentmihalyi, the University of Chicago psychologist and author of *Flow: The Psychology of Optimal Experience.*

But supposing that nanotechnology was in fact possible, its psychological effects, he thought, would not be for the better.

"It's going to be a very depressing state of affairs. Because obviously people get very depressed unless they can do things for which they feel challenged."

Happiness, he said, arose not from mindless leisure activities but from confronting and surmounting challenges. Presented with no obstacles, the mind was left with nothing to engage it, and wandered off into boredom, anxiety, or worse.

"That's usually what happens to people who retire, for instance. After being productive and putting all their energies into some kind of a job or profession and then suddenly they can't do it anymore and don't do anything — then unless they find a substitute that's equally challenging, they suffer."

The substitutes that Csikszentmihalyi proposed, however, had a distinctly *Time Machine* ring to them. "We would have to retrain ourselves to be concerned about social, emotional issues, maybe like the natives in the South Sea islands, whose energies are devoted more to interpersonal and aesthetic issues than to being good providers, instrumental producers. They prize interpersonal relations and they know how to get along; they know how to figure out what each other's needs are, and they have rituals of hospitality and rituals of recognition. They spend a lot of time singing and acting and dancing, which are skills that we have lost, and that could be rewarding."

Singing, acting, dancing. Were *they* the wave of the future — the original Stone Age entertainments coming around again for a major reprise?

"We'd have to find some new forms of expression and achievement," he said. "Otherwise people would just curl up and wither away."

Or otherwise they'd make trouble.

"You'd have a hell of a lot of time on your hands," said Clifford Geertz, anthropologist at the Institute for Advanced Study. "What you'd do with it, my sense is that you'd get into mischief, you'd get into other people's material goods — just because they were *theirs*."

"In a sense it's already happened," said Garrett Hardin, the evolutionary biologist. "Look, we have ten million unemployed; the only thing that keeps us from going crazy is the fact that we have television to divert these people. If we didn't have television I think we'd be in a great deal of trouble. We idle people and then we're surprised when they cause trouble, as in the Los Angeles riots."

But what if, after nanotechnology, the masses were supplied with all the material necessities of life?

"Most of the people involved in the Los Angeles riots *had* all the necessities of life," he said. "They've got the necessities, what they don't have is an interest in life. We deprive them of work. Basically I think activity — I won't say 'work' — *activity* is the primary requirement for human existence.

"Worthwhile activity I think is the key," he added. "So I don't think trying to strive for a world in which machines have replaced human beings is worth the effort. It would be dreadful."

"Never mind nanotechnology," said Mary Catherine Bateson, cultural anthropologist, and daughter of Margaret Mead and Gregory Bateson. "We're already in a situation where large numbers of people are becoming technologically unemployed and being replaced by robots and computers and what have you, so why are we still behaving as if the problem to solve, for human betterment, were increased productivity? Maybe the problem to solve is increased fulfillment.

"It's important for people to feel they're productive, that they *do* something, that they create value. Look at what happened to middle- and upper-middle-class women in this country in the late nineteenth, early twentieth century. What happened to a great many of them was severe depression, because — one hypothesis — is because they didn't have meaningful work to do. They had servants and they were the decorative property of their husbands."

One could extract the same lesson, just possibly, even from the animal kingdom. In zoos, for example, where animals had their physical needs taken care of, many of them spent their days pacing back and forth in the cage, hour after hour. Some zookeepers took to hiding the animal's food, making it harder to get to, just to present the beast with a challenge of some sort — something to do, a project. Furthermore, whether or not animals ever got "bored" in the wild, species living in plush environments often overpopulated themselves into decline, overrunning the carrying capacity of the habitat and then suffering a major die-out. None of this boded well for human beings glutted with luxuries from the nano realm.

But that, as usual, was only one side of the story. The other side was that at least one species had all of their physical needs taken care of and yet made out perfectly well: cats.

"Cats!" said Garrett Hardin. "They have it made! We take care of them!"

Indeed, there were some fifty-six million cats in the United States alone, and most of them could not be said to be leading miserable lives, or to be suffering any massive extinctions.

"But wild cats have a fuller life," said Mary Midgley, the British philosopher. "I am partial to domestic cats myself, but they don't have as easy a life as you think. Living acceptably with humans can be quite hard work; many fail and are put down."

Then was too much affluence a bad thing?

"Too much affluence is not a worry I've had in the contemporary world," said Peter D. Kramer, psychiatrist, and author of *Listening to Prozac*, the mid-nineties pharmaceutical-of-the-century. "The burden of poverty and need is so great that it just seems like such a long way to a society in which there are no have-nots."

Well, but wouldn't the average person go crazy, after nanotechnology, with nothing to do amid all the abundance?

"I can't imagine that," said Kramer. "There are many productive rich people. I would like to see, in my own life, the effect of enormous affluence on my productivity. It's a risk I'd be willing to take.

"There are a range of possibilities," he said. "It certainly sounds like a society that would have an enormous premium on creativity, but I can imagine a very hedonic, hedonistic society. I can imagine a very cooperative society in which there's no need for conflict, where everything's cooperative, there's great community spirit, there's rather little need for all the negative human interactions that you have to guard against in modern life. What would the purpose of theft be, and so on?"

And even if life after nanotechnology was equivalent to being retired, retirement was not necessarily the bad deal it was often cracked up to be, said Kramer. "There are some people who are very contented in retirement — other than for the problem of aging. The problem is not enough *healthy* retirement. There's a psychiatrist in Rhode Island who's just won the lottery. I don't think it's ruined his life at all."

Still, it was hard to imagine the majority of Americans, a nation of workaholics, going off to have fun for the rest of their lives while letting the nanomachines do all of the work.

"On the one hand it sounds wonderful," said Nicholas Wolterstorff, the Yale philosopher. "But there's an old Calvinist streak in me that says one or another flaw's gonna turn up."

"I tend to feel, I have a sort of nagging suspicion, that something will go wrong, that there has to be some unforeseen glitch in this vision of the future," said William McNeill, the University of Chicago historian.

Indeed, there was something unnerving, something . . . *unwholesome* about the prospect of turning the world's work over to a bunch of invisible machines. There was the impression that this was flirting with disaster, that even if it worked out in the technical sense, the nanomachines performing as advertised, nevertheless the human race did not really deserve such a thing, such an unearned boon. Lazing around in the sunshine, after all, was sloth, one of the seven deadly sins. If you were to live like that — even just temporarily, as an experiment — the odds were that something big was bound to go wrong, sooner or later. There was the impression that nanotechnology was too good to be true, that the apparent good fortune it promised was really an evil in disguise, that in the end the piper would have to be paid.

In the popular consciousness, such pessimism seemed to be an article of faith. There were all sorts of canned sermons, parables, and cautionary tales about people who got their three wishes and were destroyed by them, that you had to be careful what you wished for because you just might get it, that "more tears were shed over answered prayers than unanswered prayers," and on and on. There were all these science-fiction stories about seeming "utopias" that on closer inspection proved to be hells. It seemed to be a national theology, a common tradition, a collective superstition that the proper response to good fortune was *worry*.

But were such attitudes anything more than folk wisdom? There was an ethic behind them, of course, the so-called Protestant work ethic, the notion that honest drudgery was right and proper, that toil was the morally fitting condition of humankind. Elbow grease was morally good; as for idleness, well . . . everybody learned all about "the devil's workshop" at age four.

But what was the relevance of the work ethic in an age when physical labor was no longer required? Such a doctrine might have been appropriate before the age of nanomachines, back in olden times when you actually had to work for a living. But when you didn't, then what was the point of the dogma that said you did? Why not simply reject the ethic and enjoy life — as they already did, supposedly, in California? Gayle Pergamit, one of Chris Peterson's business partners, had a friend out there who claimed that the sole and true purpose of nanotechnology was "improved windsurfing equipment." In the generation following the nano revolution, then, perhaps nobody would give a moment's thought to the ancient and outmoded "work ethic."

Unless, of course, the work ethic was a fixed part of human nature.

"People even work at their leisure!" said Mary Bateson. "There's a basic human desire to feel you're achieving something, whether you're keeping golf scores or doing your gardening."

"Happiness does *not* increase with labor-saving, any more than it does with increased eating, beyond a certain middle-point," said Mary Midgley, the philosopher. "There would be intolerable effects from so much leisure. 'Recreation and entertainment' is quite inadequate."

"Happiness is not found in amusement," said Aristotle, back in early times — even before the moon landing. "It would be absurd to maintain that the end of man is amusement and that men work and suffer all their life for the sake of amusement. For, in short, we choose everything for the sake of something else, except happiness, since happiness is the end of man. So to be serious and work hard for the sake of amusement appears foolish and very childish, but to amuse oneself for the sake of serious work seems to be right. For amusement is like relaxation, and we need relaxation since we cannot keep on working hard continuously. Thus amusement is not the end, for it is chosen for the sake of serious activity."

Drexler's own life seemed to be a case in point. He squeezed in vacations when he could, which was to say, when he could no longer avoid them. Many's the time that Chris took off with her mother, both of them unwinding and seeing the world, while Eric stayed home and worked on a book, paper, talk, or a simulation of some grand new molecular structure. Indeed, the prospect of his spending any large amounts of time on amusements, or wallowing in such South Sea-island pastimes as dance and song, well, that was just about the most science-fictional vision you could possibly hope to associate with the name Eric Drexler.

All of which led to the question of how, after nanotechnology, the basic human need to do work, to create value, to achieve, would be satisfied. Were we going to palm off elbow grease to the nanomachines . . . only to be rewarded by making ourselves miserable? Would the irony be that work was really the good stuff of life, something that existence would be pointless without?

Was nanotechnology really just one more Faustian bargain, another apparent bonanza that would turn into a mental, moral, and physical booby trap? Was the whole nano dream some sort of cosmic-metaphysical joke?

Plausible as all of that was, there were yet a few problems with it. For one thing, the idea that people would be suicidal over no longer having to work for a living, well, that was a bit strained on the face of it. Wouldn't they be at least *slightly* relieved? After all, if they wanted to keep on working for a living, there was nothing in nanotechnology to stop them. People could do all the work they wanted to, they just wouldn't be *compelled* to — not by external reality at any rate.

The fact was that plenty of "ordinary" jobs would still be around for people to do, even in the nano age: there'd be cops, reporters, lawyers, restaurant chefs, waiters, judges, senators, writers, marriage counselors, mathematicians. Nanomachines, talented as they were, weren't going to be masters of every specialty.

Then, too, the notion of what counted as "work" would be redefined in the nano age, as it often had been in the past.

"When housework was mechanized, standards rose," said Mary Bateson. "Our ancestors didn't change the sheets twice a week, more like twice a year, probably. Back in the days when women wove their own fabric — and maybe even spun the yarn and made their own candles — the standards for what a house should look like were one heck of a lot lower than they are now. In that area, what has happened might be regarded as the creation of 'busywork,' by changing the standards.

"The identical activity can be turned into work or into leisure by being packaged differently," she added. Gardening, for example, was essentially the same activity as farming — but in one case it was "recreation," and in the other, "work." Nanotechnology would allow you to *choose* what was work, instead of having brute nature foist that work upon you.

Beyond that was the fact that even after nanotechnology was perfected, even after it had become widespread and freely available, not everyone would take equal advantage of it. Some would use it only selectively and in rare cases. Whereas few might decline nanotechnology's anti-aging benefits or its capacity for curing disease, not everyone was going to want to get their food from a "meat machine" or to live in a mock-wood, nanomachined house. Why not live in a *real* house, made out of *real* wood, built by carpenters, instead of one formed in ten minutes out of some nameless and faceless Utility Fog?

Indeed, you had to expect the usual revolt — how could it fail to happen? — a nano-backlash, consisting of a preference for the hand-made over the machine-made, a taste for the natural and the "real" as opposed to the manufactured and the "artificial." You might even foresee the creation of "Nanotechnology-Free Zones," or indeed whole communities, peopled by craftspeople, artisans, artists, and the like, who allowed no trace of Drexler's engines to defile their lives and living spaces.

"Some people may even choose to live as we do today," Drexler himself had written in *Engines of Creation.* "With traffic noise, smells, and danger; with pitted teeth and whining drills; with aching joints and sagging skin; with joys offset by fear, toil, and approaching death. But unless they were brainwashed to obliterate their knowledge of better choices, how many people would willingly resign themselves to such lives? Perhaps a few."

Perhaps even more than a few. Over and above the sheer physical issue of there being enough necessities, and even luxuries, for an entire population over an extended span of life, there were independent "lifestyle" issues to think about — choices of culture, class, mode, and status. When absolutely every last person in the neighborhood could produce their own Hope diamond in the kitchen — in Drexler's handy little "exemplar manufacturing system" — how much more stylish it would be, how much more . . . democratic, to wear, instead, a piece of hand-made wrought iron.

How much more tasteful, in fact, not to flaunt any such thing as "accessories" at all — especially such vulgar ostentations as diamond and sapphire, which is what rocket engines were now being made out of. There was a whole matter of aesthetics involved.

Often enough it was the surpassingly primitive — the native, the simple, the basic and plain — that was the sign of genuine taste and refinement, as in the case of the suburban American fireplace, which, in the late twentieth century, was the ultimate piece of home furnishing. It was far more important, status-wise, than the heat pump, than the oil-fired backup system, than the micro-climate-controlled wine cellar. All of them paled, they were nothing, in comparison to the *fireplace,* which produced, in the typical yuppie subdivision, a pall of gas and ash that hung overhead like Los Angeles smog, this big, layered smoke ring, this manifestation of "natural" heat, rising from the in-home hearth — just like in the smudged caves at Lascaux, just like in the Stone Age.

It was a basic error, apparently, to think that all would be nano in the nano age. There was no one way nanotechnology was going to play itself out, no one way the future was going to be.

* * *

As to how it would in fact work out, in detail, that was something that nobody could know until it actually happened. In the end, the case could be made that what nanotechnology meant for the human species — whether it was a godsend or a moral disaster — was not an issue that could be decided in advance. Indeed, it might not be decidable even afterward. Since when were social questions ever "decided" in any true sense anyway? After all, they weren't like scientific questions, which let you immediately run to nature, or to experiment, or to computer simulation, for verification or disproof of a given answer. In science there was an objective decision procedure to handle every question: that's what made science "scientific."

But even within science itself there were exceptions to the rule. In mathematics, for example, some questions were said to be "formally undecidable," which meant that they were not susceptible to resolution by any known or imaginable method. Such problems, it was true, tended to be on the arcane side, such as the trick theorem of Kurt Gödel which said that, within arithmetic, it wasn't possible to prove a proposition roughly analogous to the statement "This statement is unprovable."

But that wasn't the only kind of irresolvability in the sciences. There was also a separate class of problems to which there were in fact answers, only the answers were unknowable in advance on account of the inherent complexity of the situation. The weather on a given day fifty years from now, for example, depended on the combined interplay of so many different mutually related factors that it was impossible to specify it beforehand. Such problems were said to be "computationally irreducible" or "intractable." There was no way of calculating the answer that was faster than just waiting around for the actual outcome.

But if that could happen in the sciences, then why not in the humanities? If it could happen in physical systems, why not in human affairs? People weren't as predictable as atoms.

"When it comes to a technology change of that kind," said Ernan McMullin, a science historian at the University of Notre

Dame, "it's so far-reaching that the estimate of what its human impact would be, would be very premature, I would think. The difficulty of assessing technologies in general, in advance, is to know what their effects are going to be. And in the case of something like this, it's still so much on the drawing board that it's awfully hard to say anything about its impact on human life."

So maybe the ultimate meaning of nanotechnology was not knowable in advance. You couldn't predict it; you simply had to let it happen. You had to *make* it work, you had to *make* it turn out for the best, rather than decide, beforehand, that it was going to be heaven on earth or hell on wheels.

Ralph Merkle had a saying about this. Merkle, Drexler's right-hand man when it came to molecular simulations, was, one might say, nanotechnology's principal optimist, the closest thing there was to a nano-edition Dr. Pangloss. Whatever the problem, there was a solution, a workaround, a way of ironing out the difficulty. In a fundamental sense, Merkle was convinced, there were no real "obstacles."

So when he traveled around giving lectures and spreading the nano dream, he'd always finish up the same way, with the same bright quotation, which he'd gotten in turn from Alan Kay. It seemed to capture just the right spirit in words. He'd project it up there on the screen, his final thought, his closing message: "The best way to predict the future is to create it."

There was nothing to say to that, no logical rejoinder.

But it was Bernard Williams, the Oxford philosopher, who, without quite meaning to, offered the most vivid picture of the nano future.

This was in a conversation with Drexler, about the perennial and challenging topic of what people would be doing in nano era, when every obvious thing had been done, or could be, by the fabled "assemblers." Drexler, never exactly a willing participant on that theme, mumbled a few terse words about "art and performance," as if, after the breakthrough, everybody would magically become a creator.

That wasn't very likely in Williams's view: "Why should anyone suppose the conditions would exist for *producing* any art? There won't *be* any! There will be Sunday painters producing imitation trees in outer space — water colors. It'll be like Sunday afternoon in the American suburbs. It's never produced any art yet; why would it be expected to now?"

Sunday afternoon in the American suburbs.

It was the absolutely perfect image — golden and awful in equal amounts. Anything was possible, nothing was necessary.

You could picture it. Roast in the oven (cholesterol-free), or maybe an all-white-meat chicken. Family and friends around the table, including grandma — although what grandma would actually be like in an epoch when everyone was perpetually young, healthy, and prime, where maybe even granddaughter looked older than grandma, . . . well, that was just one of the novelties, one of the necessary strangenesses, one of Mark Miller's "things that are gonna be really different," after the nano breakthrough.

On the table a bottle of Dom Pérignon 1964 — produced just minutes ago in the handy home manufacturing system.

A toast to Dr. Drexler for making *that* possible!

The young ones — of the back-to-nature crowd — bringing their own bottle of wine, *natural* wine: "Real grapes. No chemicals."

The fireplace, of course, in full blaze . . .

Well, who'd not want to be there, just to see it all unfold, the Great Nano Adventure.

"I don't know that it will be utopia," Drexler told Bernard Williams, finally. "I don't claim that it will be universal misery. I claim that it's beyond my understanding, and the best that we can do is try to avoid the clear-cut dangers that we do understand today, so that people can have a chance to work out the future."

Which, in the end, seemed reasonable enough.

Who could disagree? Who'd say no? Even with all its unknowns, even with all its perils and risks, who'd say no to nano?

Selected Sources

Prologue: Mr. Nano Comes to Washington

Jeremiah, David E. Untitled speech as given by Admiral David E. Jeremiah, USN, Vice Chairman, Joint Chiefs of Staff, to the American Institute of Aeronautics and Astronautics Convention at the Naval Training Center, San Diego, Calif. February 11, 1992.

Pollack, Andrew. "Atom by Atom, Scientists Build 'Invisible' Machines of the Future." *New York Times,* November 26, 1991.

US Congress. Senate. Committee on Commerce, Science, and Transportation. *New Technologies for a Sustainable World.* Hearing before the Subcommittee on Science, Technology, and Space, June 26, 1992. 102d Cong., 2d sess., 1992. S. Hrg. 102-967.

1. The *kT* Irony

Brown, Robert. "A brief account of microscopical observations made in the months of June, July, and August, 1827, on the particles contained in the pollen of plants; and on the general existence of active molecules in organic and inorganic bodies." *Philosophical Magazine* 4 (1828): 161. Also, "Additional remarks on active molecules." *Philosophical Magazine* 6 (1829): 161. Both papers reprinted in *The Miscellaneous Botanical Works of Robert Brown,* edited by J. J. Bennett. 2 vols. London, 1866–67.

Brush, Stephen G. *The Kind of Motion We Call Heat: A History of the Kinetic Theory of Gases in the 19th Century.* 2 vols. Amsterdam: North-Holland, 1976.

Burgess, Jeremy, Michael Martin, and Rosemary Taylor. *Under the Microscope.* Cambridge: Cambridge University Press, 1990.

Morrison, Philip, Phylis Morrison, and the Office of Charles and Ray Eames. *Powers of Ten.* New York: Scientific American Library, 1982.

Nye, Mary Jo. *Molecular Reality: A Perspective on the Scientific Work of Jean Perrin.* New York: Elsevier, 1972.

Schrödinger, Erwin. *What Is Life? The Physical Aspect of the Living Cell.* New York: Macmillan, 1946.

2. The Old Technology

Laue, Max von. "Concerning the Detection of X-ray Interferences." Nobel lecture, November 12, 1915. In *The Nobel Prize Winners,* edited by Frank N. Magill, vol. 1 (1901–1926). Salem Press, 1989.

Müller, Erwin W. "Atoms Visualized." *Scientific American* 196 (June 1957): 113.

———. "A New Microscope." *Scientific American* 186 (May 1952): 58.

———. "Resolution of the Atomic Structure of a Metal Surface by the Field Ion Microscope." *Journal of Applied Physics* 27 (May 1956): 474.

———. "Study of Atomic Structure of Metal Surfaces in the Field Ion Microscope." *Journal of Applied Physics* 28 (January 1957): 1.

3. The World-Class Auto-da-fé

Drexler, Kim Eric. "Design of a High Performance Solar Sail System." Master's thesis. MIT, 1979.

Krumhansl, James A., and Yoh-Han Pao. "Microscience: An Overview." *Physics Today 32* (November 1979): 25.

4. "Feynman Was Robbed"

Feynman, Richard P. "There's Plenty of Room at the Bottom." *Engineering and Science* 23 (February 1960): 22. Reprinted in *Miniaturization,* edited by H. D. Gilbert. New York: Reinhold, 1961.

Hippel, Arthur R. von. "Molecular Designing of Materials." *Science* 138 (October 12, 1962): 91.

———, ed. *Molecular Science and Molecular Engineering.* Cambridge: MIT Press, and New York: John Wiley, 1959.

"Protein Molecules Interface to Microcircuitry." *Semiconductor International* 3 (May 1980): 10.

5. Eternity and Clouds

Drexler, K. Eric. "Space Colony Supply from Asteroidal Materials." In *Space Manufacturing Facilities (Space Colonies)*, edited by Jerry Grey. New York: American Institute of Aeronautics and Astronautics, 1977.

Meadows, Donella H., Dennis L. Meadows, Jørgen Randers, and William W. Behrens III. *The Limits to Growth: A Report for the Club of Rome's Project on the Predicament of Mankind.* New York: Universe Books, 1972.

6. Richard Comes to Chris and Eric's

Drexler, K. Eric. "Molecular Engineering: An Approach to the Development of General Capabilities for Molecular Manipulation." *Proceedings of the National Academy of Sciences* 78 (September 1981): 5275.

Hofstadter, Douglas R. "Self-Referential Sentences: A Follow-Up." In *Metamagical Themas,* edited by Douglas R. Hofstadter. New York: Basic Books, 1985.

Nelson, Ted. *Computer Lib / Dream Machines.* Self-published, 1974. Reprint. Redmond, Wash.: Tempus Books, 1987.

7. Brother Eric's Nanotech Revival

Gold, Michael. *A Conspiracy of Cells: One Woman's Immortal Legacy and the Medical Scandal It Caused.* Albany: State University of New York Press, 1986.

Pabo, Carl. "Designing Proteins and Peptides." *Nature* 301 (January 20, 1983): 200.

Ponder, Jay W., and Frederic M. Richards. "Tertiary Templates for Proteins." *Journal of Molecular Biology* 193 (1987): 775.

Ulmer, Kevin M. "Protein Engineering." *Science* 219 (February 11, 1983): 666.

8. Tiny *Tale* Gets Grand

Dietrich, J. "Tiny *Tale* Gets Grand." *Engineering and Science* 49 (January 1986): 24.

Drexler, K. Eric. *Engines of Creation.* New York: Doubleday, 1986.

Feynman, Richard P. "Infinitesimal Machines." Talk at the Jet Propulsion Laboratory, California Institute of Technology, 1983. Caltech Archives.

———. "Tiny Machines." Talk at the Esalen Institute, 1984. Mill Valley, Calif.: Sound Photosynthesis, 1984.

Heinlein, Robert A. [Anson MacDonald, pseud.]. "Waldo." *Astounding Science Fiction,* August 1942. Reprinted in *Three by Heinlein: The Puppet Masters, Waldo, and Magic, Inc.* New York: Doubleday, 1965.

9. Astrid and Priscilla

Baeyer, Hans Christian von. *Taming the Atom.* New York: Random House, 1992.

Dehmelt, Hans. "Experiments on the Structure of an Individual Elementary Particle." *Science* 247 (February 2, 1990): 539.

———. "Experiments with an Isolated Subatomic Particle at Rest" (Nobel Lecture). *Angewandte Chemie International Edition in English* 29 (July 1990): 734.

———. "A Single Atomic Particle Forever Floating at Rest in Free Space: New Value for Electron Radius." *Physica Scripta* T22 (1988): 102.

Drexler, K. Eric. "Nanomachinery: Atomically Precise Gears and Bearings." In *IEEE Micro Robots and Teleoperators Workshop.* Hyannis, Mass., 1987. IEEE, cat. no. 87TH0204-8.

10. Monotony, Hate, and Utopia

Dewdney, A. K. "Nanotechnology: Wherein Molecular Computers Control Tiny Circulatory Submarines." *Scientific American* 257 (January 1988): 100.

Drexler, K. Eric. "Rod Logic and Thermal Noise in the Mechanical Nanocomputer." In *Molecular Electronic Devices,* edited by Forrest L. Carter, Ronald E. Siatkowski, and Hank Wohltjen. Amsterdam: North-Holland, 1988.

Friedman, David. "Economic Consequences of Nanotechnology." Talk at MIT Nanotechnology Study Group Symposium, January 1987.

MacGillivray, Jeffrey C. "The Economics of Rapidly Changing Technology." Parts 1, 2. *Foresight Update,* no. 7 (1989): 5; no. 8 (1990): 5.

Nilsson, Nils J. "Artificial Intelligence, Employment, and Income." *The AI Magazine,* Summer 1984, 5.

11. "Three Little Gears"

Becker, R. S., J. A. Golovchenko, and B. S. Swartzentruber. "Atomic-Scale Surface Modifications Using a Tunnelling Microscope." *Nature* 325 (January 29, 1987): 419.

Binnig, Gerd, and Heinrich Rohrer. "The Scanning Tunneling Microscope." *Scientific American* 253 (August 1985): 50.

DeGrado, W. F., L. Regan, and S. P. Ho. "The Design of a Four-Helix Bundle Protein." *Cold Spring Harbor Symposia on Quantitative Biology* 52 (1987): 521.

DeGrado, William F., and Lynne Regan. "Characterization of a Helical Protein Designed from First Principles." *Science* 241 (August 19, 1988): 976.

Foster, J. S., J. E. Frommer, and P. C. Arnett. "Molecular Manipulation Using a Tunnelling Microscope." *Nature* 331 (January 28, 1988): 324.

"Microscopy by Vacuum Tunneling." *Physics Today* 35 (April 1982): 21.

Pollack, Andrew. "New Generation of Tiny Motors Challenges Science to Find Uses." *New York Times,* July 26, 1988.

Quate, Calvin F. "Vacuum Tunneling: A New Technique for Microscopy." *Physics Today* 39 (August 1986): 26.

12. Captain Future

Garfinkel, Simson. "Critique of Nanotechnology." *Whole Earth Review*, Summer 1990, 104.

Gustavson, Todd Lyndell. "Design of a Simplified Scanning Tunneling Microscope." International Science and Engineering Technical Paper. 1989.

Lewis, James B., and John L. Quel (eds.). *Nanocon Proceedings.* Seattle, Wash.: Nanocon, 1989.

Policy Research Project on Anticipating the Effects of New Technologies. *Assessing Molecular and Atomic Scale Technologies (MAST).* Austin: University of Texas, 1989.

13. "Are Molecules Sacred?"

Bugg, Charles E., William M. Carson, and John A. Montgomery. "Drugs by Design." *Scientific American* 269 (December 1993): 92.

Crandall, BC, and James Lewis, eds. *Nanotechnology: Research and Perspectives. Papers from the First Foresight Conference on Nanotechnology.* Cambridge: MIT Press, 1992.

Eigler, D. M., C. P. Lutz, and W. E. Rudge. "An Atomic Switch Realized with the Scanning Tunnelling Microscope." *Nature* 352 (August 15, 1991): 600.

Eigler, D. M., and E. K. Schweizer. "Positioning Single Atoms with a Scanning Tunnelling Microscope." *Nature* 344 (April 5, 1990): 524.

"Engineering a Small World: From Atomic Manipulation to Microfabrication." *Science* 254 (November 29, 1991): 1300.

Hapgood, Fred. *Up the Infinite Corridor: MIT and the Technical Imagination.* Reading, Mass.: Addison-Wesley, 1993.

Meadows, Donella H., Dennis L. Meadows, and Jørgen Randers. *Beyond the Limits: Confronting Global Collapse, Envisioning a Sustainable Future.* Post Mills, Vt.: Chelsea Green, 1992.

Quate, C. F. "Switch to Atom Control." *Nature* 352 (August 15, 1991): 571.

14. The Greek Chorus of Woe

Drexler, K. Eric. *Nanosystems: Molecular Machinery, Manufacturing, and Computation.* New York: John Wiley and Sons, 1992.

Drexler, K. Eric, and Chris Peterson with Gayle Pergamit. *Unbounding the Future: The Nanotechnology Revolution.* New York: William Morrow, 1991.

Mislow, Kurt. "Molecular Machinery in Organic Chemistry." *Chemtracts — Organic Chemistry* 2 (1989): 151.

15. "Good Luck Stopping It"

Curl, Robert F., and Richard E. Smalley. "Fullerenes." *Scientific American* 265 (October 1991): 54.

Dagani, Ron. "Scientists Rapidly Breaking New Ground for Nanoscale Technology." *Chemical and Engineering News* (March 22, 1993): 20.

Elmer-DeWitt, Philip. "The Genetic Revolution." *Time*, January 17, 1994, 46.

Smalley, Richard E. "Smalley on Nanotechnology." *Rice News* (November 11, 1993): 8.

Yam, Philip. "The All-Star of Buckyball." *Scientific American* 269 (September 1993): 46.

16. Sunday Afternoon in the American Suburbs

Wells, H. G. *The Time Machine*. 1895. Reprinted in *The Science Fiction Hall of Fame*, edited by Ben Bova, vol. 2A. New York: Doubleday, 1973.

Acknowledgments

The author would like to thank all those who provided help in the preparation of this book. For source materials, documents, and contacts: Ivan Amato, Faustin Bray, Steve Bridge, BC Crandall, Kevin Finneran, Josh Fishman, Richard B. Gerber, Robin Gill, Evan Hadingham, John Storrs Hall, David Lewin, Neil McAleer, Masako Ohnuki, Dennis Overbye, Gary Pastore, John Pitts, Mark Plus, Boyce Rensberger, and Conrad Schneiker. For nano-news and general gossip: Fred Hapgood, Keith Henson, and Gary Hudson.

Particular thanks to James Gleick for a critical piece of advice.

For reading and checking all or part of the manuscript, many thanks to Eric Drexler and Chris Peterson, and to Hans C. von Baeyer, Larry Marschall, and Mary Jo Nye.

I am indebted to several current and former faculty members, librarians, staff, and officers of Western Maryland College — specifically: Robert H. Chambers, Mark Collier, Ed Holthause, Jennie Mingolelli, Louise Paquin, David Seligman, and Cheri Smith.

Special thanks to Teresa Ehling of MIT Press for a prepublication copy of *Nanotechnology: Research and Perspectives*, furnished to the author on diskette; to Virginia A. Story for tape transcription; to my wife, Pamela Regis, for uninterrupted moral support and other essential help; to my literary agent, Jean V. Naggar, for rescuing me from a succession of hair-raising life experiences; and to my editor, Fredrica S. Friedman, for advice and criticism through several drafts of the manuscript.

Above all, I am indebted to Eric Drexler and Chris Peterson, without whose trust and support this book could not have been written. They cooperated fully from beginning to end, never denied a request, neither sought nor received manuscript approval, and even at the most doubtful moments displayed an inordinate confidence in the author.

Index

Page numbers in *italics* refer to illustrations.